AF679502

Mitochondrial Eve and Y-Chromosomal Adam: A Defined Redux of Human Evolution

Subir Ranjan Kundu (Author)

Jessica English (Managing Editor)

Academica Press
Washington~London

Library of Congress Cataloging-in-Publication Data

Names: Kundu, Subir Ranjan (author)
Title: Mitochondrial eve and y-chromosomal adam : a defined redux of human evolution | Subir Ranjan Kundu
Description: Washington : Academica Press, 2022. | Includes references.
Identifiers: LCCN 2022944883 | ISBN 9781680537024 (hardcover) | 9781680537031 (paperback) | 9781680534399 (e-book)

Copyright 2022 Subir Ranjan Kundu

Dedication

To my aunt Dr. Shila Kundu Poddar, Anthropological Research, Ex-Curator, Indian Museum, Kolkata, India. The unceasing flow of inspiration, hands-on guidance, and her endless psychological support helped me a lot in boosting up my enthusiasm to carry on my journey of life through the unavoidable, and state of seclusion in life, unless writing of this book would have been remain unfinished.

Contents

Foreword

This book, entitled **"Mitochondrial Eve and Y-Chromosomal Adam: A Defined Redux of Human Evolution,"** authored by Dr. Subir Ranjan Kundu is an earnest attempt to find the grassroots of human evolution along with a review of the biblical narrative of the prehistoric emergence of Adam and Eve as the universal ancestral lineage of humanity on planet Earth. I realize that this book will create a great deal of curiosity amongst the readers.

It has never been easy to challenge the consensus because the System – of any kind, in any context – will try to preserve the status quo, by all means possible. Having spent a lot of period in studying evolutionary biology and molecular genetics of higher organisms, especially the eukaryotes, and contributing numerous articles in the international journals with five books to my credit, and I feel obliged to share my knowledge, analyses, and conclusions. Hopefully, this study will raise the level of awareness among the science enthusiasts, researchers especially those interested in understanding the domain of evolutionary biology, especially the anthropogenic origin of "Sapiens," its evolution and diversification, and initiate the discussion that, in turn, may entail major changes in evolutionary biology.

Evolution has neither been recognized as an ideal process nor a perfect set of designs to ascertain the smooth progress of life in the living world, rather it should be regarded as an optimal safeguard that ensures the survival of a living entity in the changing environment. Literally, "Evolution" appears to be a dynamic state of a living entity, clearly demonstrates its change of state, in the background of Spatio-temporal changes, and complied with the fitness of "natural selection." I am sure, the book would fascinate its readers by holding their attention until the readers complete reading from the beginning to the end.

It's a great opportunity for me to read the book authored by Dr. Subir Ranjan Kundu to read it with a great deal of interest. In the initial part of the book, the author has critically reviewed the hypotheses and theories and synthesized the secondary data with substantial pieces of evidence to support the fossil evidence to interpret the findings into facts to decrypt the history origin and evolution of "Sapiens" on Earth, buried under the stack of evidence, theories, and hypotheses. The vertical inheritance of evolutionary adaptations from the ancestors to its successors reflecting the existence of hereditary connections and their expression in the descendants, vertically acquired from their forefathers through genetic recombination and the expression of archaic traits in the ancestral lineages have gradually replaced by the unique traits in the derived group of organisms, that helped them to sustain and evolve further in the ever-changing environment.

At times, the theories and hypothesis presented might seem overwhelming. I hope that you won't be easily discouraged. Even if the content of a given chapter is hard to understand, the scientific evidence presented, the citations from original documents, conclusions drawn, and recommendations made can be easily comprehended. One of the important features of this book is that it does not have a textbook structure when the chapters, in order to be understood, need to be read in the sequence given. In fact, you can start the journey from any chapter, based on your interests, tastes, and preferences. I do not expect everybody in the scientific community to agree with the content and ideas put forth in this book. But I do hope that the information and knowledge presented will trigger the thought process of young researchers studying evolutionary biology and molecular genetics.

This book comprises of 12 chapters and is an unbiased portrayal of unique narration with critical reviews to witness a major evolutionary transition, humans, to depict the ultimate journey of evolution of "Sapiens" on planet Earth in the current regime of Anthropocene.

The book raises the curiosity of the reader as the story takes them on a unique journey of evolution in quest for the Y-Chromosomal MRCA Adam, Mitochondrial Eve & proliferation of Humankind contradicting it with genetic diversity. The work mainly evolves round the origin, gradual evolutionary development and transition of Sapiens on earth supported by various hypothesis and evidences. It explores the latest research in anthropogenic evolution, of the hominoids and hominine in the spatiotemporal backdrop of the environmental dynamics. It is an exploration of the major transitional links in the biological world starting from the archaic hominoids (like Big Apes) and archaic hominines (like Homo erectus, Neanderthals, etc.) and ending with the phylogenic footprints of the extant (like Chimpanzees, Gorillas, etc.),

The book also encompasses references of the work done by a group of molecular geneticists, genomicists, and evolutionary biologists who claimed that all humans originated from a primordial ancestors /couple recognized as prehistoric universal mother and father relating with the biblical narrative of the prehistoric emergence of Adam and Eve as the universal ancestral lineage of humanity on planet Earth, this hypothesis being challenged by another group of molecular geneticists, genomicists, who stated that there was huge genetic diversity in humans as a result of mutations or errors in DNA replication and denied the fact that the primordial couple had contributed in genetic diversity or had any inherent genetic diversity. They claimed that the genetic diversity observed now in the entire human population is incompatible with the hypothesis of origin of humans from a prehistoric couple, popularly known as Y-chromosomal Adam (genetic father) and Mitochondrial Eve (genetic mother).

In this context an effort was made to study the "single-couple origin model of humanity" in which the allele frequency spectra, and linkage equilibrium statics from contemporary genetic data emphasized on the probability of a "primordial-couple" origin of humanity which occurred in the African continent, around 100,000 years ago, before archaic predecessors .As a result a consensus was reached wherein it was accepted that there must have been a small-sized archaic human population of one to few thousand individuals, which existed around 100,000 years ago, and contributed to the present level of genetic diversity in modern humans species.

I hope this book is widely read, studied and analyzed in order to understand the evolutionary biology. We need to understand and interpret the various hypothesis put forth and start benefiting from the knowledge base created by the scientists. My best compliments to Dr. Subir Ranjan Kundu, who has realized that it's an exigency of time to serve the "needs" of the readers of academic and non-academic background to grasp the complex conceptions of biological evolution, related to our ("Sapiens") history of origin and evolution on planet Earth.

Dr. Varsha V. Andhare
Associate Professor & Research Supervisor
Department of Zoology
The Institute of Science, Mumbai
Maharashtra
India
E-mail:varshandhare@yahoo.com

Preface

"Mitochondrial Eve and Y-Chromosomal Adam: A Defined Redux of Human Evolution," authored by Dr. Subir Ranjan Kundu, has initially resorted with theoretical ideas, interpretations, and paleontological evidence to narrate the origin and evolutionary story of "Sapiens," passed through the transitional stages of archaic human species, involved in the evolutionary pilgrimage, once started in form of the great apes and gradually evolved in the form of "Sapiens," after traversing a long evolutionary trail. Eventually, the author has shifted his area of interest from paleontological analysis to the molecular investigations of the DNA footprints of primates including great apes, archaic humans, and anatomically modern human beings, to stretch out the missing links in between the evolutionary milestones to define and redefine the evolutionary progress of life transcended through archaic humans to be evolved as "Sapiens," the ultimate mammal in the regime of Anthropocene on planet Earth. The origin and evolution of Humans have always been remained a source of debate between the creationists and evolutionists, in terms of recognizing the results of such researches on biological evolution, and its credible interpretation of the evolutionists who upheld the origin and evolution of "Sapiens" resulting from great apes in course of the gradual evolutionary progress of life. On the contrary to the evolutionary origin of "Sapiens," the Book of Genesis, in the Bible, has often been referred to as a very poetic, even the mythological version of the special creation of humans, believed to be created out of Adam and Eve, the primordial couple supposedly lived in the Garden of Eden and this special creation of human beings has supposedly taken place by the grace of God and a large number of people hardly believes that this could ever have happened. In this book, the author has made an earnest effort to review include the religious interpretation of the emergence of human beings on Earth, according to the principle of the special creation in "Genesis," before understanding the scientific principles of biological evolution of "Sapiens" on our planet. Finally, the author has engaged in the contemporary analysis of interpretations of the molecular and evolutionary geneticists and genomicists, engaged in search of molecular footprints of "Sapiens" for the last four decades. And, it has prompted the author to present a detailed illustrations on the matrilineal inheritance of mitochondrial DNA (represented by mitochondrial Eve as primordial mother), patrilineal inheritance of Y-chromosomal DNA (represented by Y-chromosomal Adam as primordial father), and presented an elaborated structural aspects of the human genome and molecular aspects of DNA footprint of "Sapiens," that would make this book as a most interesting piece of reference to the heterogeneous readers, graduate and post-graduate students, research scientists and knowledge-seeking persons, looking for a brief understanding on the origin and biological evolution of humans on Earth in the backdrop of molecular phylogenetics.

Acknowledgments

I express my deep sense of gratitude and indebtedness to Dr. Pradipta Kumar Sarangi, Lecturer, Department of Zoology, Indira Gandhi Memorial College, Odisha, India; Dr. Parth Pratim Dhar, Principal/Professor, Sir Ashutosh Govt. College, Chittagong, Bangladesh; Pierre de Chabbannes, Scientific Advisor, National Geographic Photo Ark, Paris, France; Dhaval Bhatt, Assistant Professor, Department of Zoology, Gyanmanjari Science College, Bhavnagar, Gujarat, India; Sileesh Mullassei, Reseracher/Freelance Science Writer, Schhol of Ocean Science & Technology, Kerala University of Fisheries and Ocean Studies, Cochin, India, and Dr. Varsha V. Andhare, Associate Professor, Department of Zoology, The Institute of Science, Maharashtra, India; for their suggestions and guidance in this present writing.

I am especially pleased to Dr. S. Sundar, Research Scientist, S.S. Research Foundation, Kallidaikurichi, Tamil Nadu, India; Ms. Jaya Sarkar, PhD Fellow, Department of Humanities & Social Sciences, Birla Institute of Technology & Science-Pilani, Hyderabad Campus, Telangana, India, who have helped me in drafting the different part of the manuscript, required for further reviewing consultations of literature, library consultation and seeking opinion of the subject matter experts etc.

I am also thankful to Susan Zadek and Sung-Min Han, Albert Campbell Library; Gail MacFayden and Eva Lew, Bloor Gladstone Library, under Toronto Public Library Network, Toronto, Canada for extending Library facilities and providing necessary help in consulting the literature required for this project to undertake.

I would like to express my gratitude to Ms. Arkatapa Basu, Trainee Journalist, Deccan Herald, Bengaluru, Karnataka, India the freelance editor, proof reader and content developer, who has gone through the text and do the needful in the final phase of literary editing of this book. I am expressing my gratitude to thank all authors, editors, publishers of books and journals, colleagues for sharing the printed and digital materials, providing valuable suggestions and tendering necessary clarifications during the last few years, selflessly helped me out to complete this task.

Chapter 1

To reach for a rainbow, love and creation, it's all about the experience of first bite of the apple.

And the Lord God commanded the man, saying, Of every tree of the garden thou mayest freely eat: But of the tree of the knowledge of good and evil, thou shalt not eat of it: for in the day that thou eatest thereof thou shalt surely die.

---- Genesis 2:16-17

In January of 1942, the Editor's office of *The Daily Telegraph*, United Kingdom (UK) received several complaints from subscribers. The complaints were that the crosswords were not challenging enough to crack, as it took only a few minutes to solve them. This made WAJ Gavin, the Chairman of the Eccentric Club in London, UK, to come up with a unique plan to crack a cryptic crossword puzzle within a 12 minutes deadline, and further announced that if anyone would successfully solve it within the deadline, a prize money of £ 100.00 would be donated to a charity. Arthur Watson, the Editor of the *Daily Telegraph* had come forward to sponsor this unique endeavor and to hold this competition at the Newsroom at Fleet Street, London, on 12th January 1947. The editorial office, was quite surprised to witness that there were four competitors that successfully met the challenge of the *Daily Telegraph*'s sponsored crossword championship, within the 12 minute deadline and in the next day's edition. The crossword puzzle was published in hopes that it would be shared among vast array of their subscribers, and could be good end of a successful organization of this cryptic crossword puzzle event (Chivers, October 10, 2014). But, it was basically the beginning of a bigger story as the successful completion and publication of this special cryptic crossword puzzle, on the 13th January 1942, "*The Daily Telegraph Crossword Time Test*" in the daily edition of

The Daily Telegraph, played a cardinal role for Allied force to win the Second World War (WWII) against Axis Power camp (Diffen.com, May 16, 2011). The contestants took part in this competition did not know that the war office intelligence and the Royal Military Intelligence in Great Britain had kept an eye on as well as being indirectly involved in the crossword puzzle event, and they also kept eyes on the successful participants of that competition. Stanley Sedgewick, one of the four successful participants of this competition had received a letter marked "Confidential" after few weeks of the competition. The letter instructed him to meet Col. Nichols, the General staff of British Intelligence, and to appear before him to take part in a discussion on a matter of national importance. In 1998, author Michael Smith has mentioned in his non-fiction novel, "*Station X: The Codebreakers of Bletchley Park*," that Sedgewick and a number of participants of that cryptic crossword competition on the 12th January, 1942, ended up appointed to working on the project "Ultra" in the "Hut-8" location at Bletchley Park, London, to crack the Enigma codes, the encoded communication extensively used by the German military (Smith, 1998). In the history of artificial intelligence, specifically in the history of computing and cryptanalysis. The invention of the Enigma machine, the mechanical encryption machine (a set of rotor machines), and its development and strategic use for communication in military intelligence, should be considered as one of the most important factors that tilted the equilibrium of power-play from Axis Powers (Italy, Germany, Japan) or Axis of Rome-Berlin-Tokyo to The Allied Powers (United Kingdom, Soviet Union, United States of America along with France, Poland, Belgium, Greece, Yugoslavia, Canada, Australia, New Zealand etc.) during WWII (1939-1945) (Diffen.com, May 16, 2011). According to a very popular Hollywood movie *The Imitation Game* (2014), which was inspired by the non-fiction literature authored by Andrew Hodges, *Alan Turing - The Enigma*, scientists have invented a mechanical encryption machine (which was made up of a key board, a plug board and a set of rotors) called "Enigma" during the wake of the First World War (from 28th July 1914 to 11th November 1918). It was intended to protect the extremely important information like diplomatic and strategic information, as well as hiding sensitive documents like intelligence reports and data on commercial intelligence etc. (Diffen.com, May 16, 2011). The structural conformation of this machine was apparently simple as the internal wiring of the plugboard of Enigma could be changed to alter the function of encryption keys, and it had a reversing drum and several lamps corresponding to specific letters. Germany, Italy and Japan started using the Enigma machine immediately after its invention during First World War, and initially it was used for commercial communication (Diffen.com, May 16, 2011). In the field of encryption-flanged, code-based communication systems, which played a key role in the progress and dominance of German military in the Allies- Axis conflict in Europe, the extensive usage, research and further development of Enigma was the key communication and strategic advantage during WWII. At the end of the Second World War, US President Dwight Eisenhower commented that breaking of the Enigma code shortened the War by at least 2 years which most likely saved millions of souls (DiSalvo, May 27, 2012).

During WWII, the German naval force, specifically, the German U-boats, overpowered the naval establishments, resources and bases of Allied forces across Europe, most of the North American naval fleets, those stationed at the Eastern coasts of United States, the Gulf of Mexico and Caribbean sea. They were in the range of German submarine attacks and the situation for Allied Power was so bad in the first quarter of 1942, where the German naval fleets, specifically U-boats sank around a hundred US naval ships along the eastern coasts of the United States (DiSalvo, May 27, 2012). The Enigma encoded message, which was the cardinal strategic support of the Axis power camp and German naval fleets, played a cardinal role in their unsurmountable progress and fearsome presence in Axis-Allied conflicts of the first half of WWII. In order to offer a strategic help to counter the progress of Axis Power, the Intelligence support of Allied forces and the British Military Intelligence had received help from a number of extremely efficient, mathematicians and cryptanalysts that engaged in decrypting the encoded messages of German naval team at "Hut 8," in Bletchley Park, WWII code breaking station, and the entire exercise had been monitored and coordinated by one of the most enigmatic mathematician of Cambridge, and Logician Alan Mathieson Turing (https://www.cia.gov/news-information/featured-story-archive/2015; CIA.gov, April 10, 2015). In order to counter the functionalities of the German improvised Enigma machine, in 1939, Turing had developed an electromagnetic instrument called "Bombe" by sharing the findings of the Polish and French cryptanalysts, which helped him and his team members to crack the encoded messages of German ciphers. It was the meticulous endeavor of Turin and his colleagues at Hut-8, who spent every minute of the three years (1941-1943) to decode the infamous German naval code "Enigma" (https://www.cia.gov/news-information/featured-story-archive/2015; CIA.gov, April 10, 2015). It was Polish mathematician and cryptanalyst, Marian Rejewski, who was successful in decoding the infamous Enigma code, developed and used by German navy in 1932, and he was the first one to design the earliest form of "Bombe," the decoding electromagnetic device, which was later used and developed in 1939 for decoding the Enigma code. During German occupation of Poland in 1939, Polish cryptanalyst Rejewski, moved with his invention to France to be a greater part of Allied power where he shared the basic design and functionalities of "Bombe" to their British counterparts, which assisted Turing to develop "Bombe" in its effective form which cracked the Enigma ciphers. However, It was a cumbersome task and a serious challenge to the cryptanalysts of the Allied power camp, as the German military intelligence had not only acquired the Enigma technology to encode or decode the messages immediately after the First World War, but their military intelligence had engaged in its research and development for next 20 years to upgrade the Enigma technology that rendered it almost impossible to crack by any conventional cryptanalyst in the world. Military historians have always contended that without the earnest endeavor of Turing and his team members at Hut-8, it was almost impossible for allied force to crack the cryptic messages of the German Naval forces in WWII, as the German Military intelligence had superior ciphering technologies in their possession, which they utilized through the Enigma machine (DiSalvo, May 27, 2012). Even the prime minister and one of the legend leaders of Allied camp during WWII, Winston Churchill had accepted the fact that without the help of "Bombe," which played a major role in locating the key locations of German submarines and German U-Boats, along the east coast of United States, and the German naval fleets being destroyed by the naval fleets of the Allied forces, which were very precise to cut the supply-lines of Britain during the War, it was most likely result in a losing the "Battle of Atlantic," and WWII as well. Once the British Military Intelligence were able to decrypt the secret messages of the German navy, they shared all important observations of project "Ultra" from Bletchley park, to the OSS (Office of Strategic Services), the precursor of CIA (Central Intelligence Agency), and the establishment of national-level counterintelligence Division in the secret intelligence branch in the United States during WWII (https://www.cia.gov

/news-information/featured-story-archive/2015; CIA.gov, April 10, 2015).

After WWII, Turing had took a personal endeavor to invent the earliest edition of modern (personal) computers and Turing was considered as the pioneer of “Artificial Intelligence” (DiSalvo, May 27, 2012). In 1950, Turing had published one of his most important papers, *Computing Machinery and Intelligence*, where he expressed his belief about the unlocking potential of the modern day computers, which was considered to be equally and precisely comparable to the human brain in terms of thought process and memory retention. In order to adjudge the enigmatic academic potential of this unique scholar, we need to make a lateral review on the chronological progress of Industrial Revolution (Muir, July 24, 2019 in https://aibusiness.com/livetiles-fifth-industrial-revolution/). We are aware that before initiation of Industrial Revolution, that there were around 80% of people that sustained their living through agricultural practices. The beginning of Industrial Revolution, which symbolically started in Great Britain with steam power in 1760, and then continued till 1870, expanded a new horizon in terms of social advancement (in terms of GDP growth and surge of urbanization) across Europe, and proven exponential industrial growth in the sectors of textile, chemical and iron (heavy metal). The second Industrial Revolution, which begun in 1870 and ended in 1914, had enabled socio-economic development of the wide array of urban settlers by pushing forward the wheel of Industrialization, particularly in the sectors of electrical, petroleum, steel and automobile industries. Overall progress has further been assisted by the extensive usage of automation in industrial sectors and remarkable progress in communication sectors, where wide-spread expansion of the network of railways and telephone played its fundamental role. The third Industrial Revolution which was overlapped with 2nd one continued till the 1960s, had been marked with invention of the microprocessor and initiated the beginning of cyber revolution by the invention of computers for use it in industries and office places. The use of new terminologies like Minimum Retail Price, Supply Chain Management, Visual Merchandising, became common terminologies in the Industrial sectors during this time. The ending of the 3rd Industrial Revolution begun with the successful change of technology transfer from analog technology to digital technology. The fourth Industrial Revolution (which has also been recognized as 4.0) which was found to be overlapping in nature with 3rd Industrial Revolution, which started a new era of Cyber Revolution, and was considered to be started in 1980-90s with the beginning of the journey of digital transformation, which acted as a connecting technology between cyber-human interface. Eventually, it started surging in its highest level from 2010 with immense progress in artificial intelligence, cloud computing, 3D and digital printing, and 3G and 4G technologies in the contemporary communication sector. Now a days, we are in the era of 5th Industrial revolution, which also overlaps in nature with 4th Industrial Revolution, where extensive proliferation of virtual workplace, Artificial Intelligence (AI) and robotics took over the work-load and decision making process of human beings, and became a different ball-game as we don't know what would be the ultimate consequence of the disappearance of the conventional 9-5 office jobs.

Let's go back to Turing's ground-breaking dissertation on "*Computing Machinery and Intelligence*," which was published in 1950, and helped him to earn the recognition as "Pioneer in Artificial Intelligence," who had foreseen and apprehended the future uncapped potential of technological revolution in the era of the 5th Industrial Revolution, and did not see it in real life, though his dream came true when he was not in this world (Turing, 1950; DiSalvo, March 27, 2012). In 1952, in a police investigation after an incident of burglary in the home of Alan Turing, his deepest secret on his sexual orientation came to the surface as he had been found engaging in homosexuality with a sex partner of a 19-year-old man. He then lived in Manchester (Britain), where the act of homosexuality had been considered a criminal offence in that time (DiSalvo, May 27, 2012). Alan Turing faced the same fate that the great poet Oscar Wilde faced once, where he had been legally prosecuted under law of "Gross Indecency" for such criminal offence, and he had been given two choices either to accept one year imprisonment, or to accept a hormonal treatment to reorient his sexual orientation. The much publicised news scoop in the British media during 1952 about the personal life story of a war hero like Turing, tarnished his public image, and being embarrassed by media stir, had confined himself to avoid any public contact, even though Turing had chosen to take the hormonal treatment, which was basically a way of physio-chemical castration. However, Turing continued a regular intake of the hormone estrogen as the part of hormonal treatment which resulted in serious side-effects on his physical and mental health, and he became a victim of another hormonal ailments during this period, known as "gynecomastia."[2] In the 1950s there was not enough media control and they would often go above and beyond moral ethics when covering stories about someone's personal life, so the follow-up news coverage about his personal life, shattered the psychological strength of Turing and he did not find any other option but to end his journey by committing suicide at the age of 42 in 1954 (Cooper, November 18, 2014; https://www.cia.gov/news-information/featured-story-archive/2015; CIA.gov, April 10, 2015). A number of people assumed that Turing did not commit suicide, rather he tried to experiment with cyanide, but most of the people considered that Turing was desperate to get rid of his life by committing suicide as he found that he would never get back the respectful status that he had once earned for his contribution during WWII. However, the official statement of British government has further corroborated the "suicide hypothesis" as the cause of death of this genius, on the basis of the examination of his blood sample and an apple that had a bite in it were found to have traces of cyanide. This rendered Alan Mathias Turing to cherish the "kiss of death," and the news had been published in 11th June 1954's edition of the Daily mail under the headline "*Death Apple by his bed*" (Cooper, November 18, 2014). In this age of cyber revolution it's not easy for us to judge whether the British legal system was appropriate during this time for the diverse group of people with a various sexual orientations, or whether the sexual orientation of some people should have been considered inappropriate at the time. Throughout this passage of time our society has gone through a number of changes including those of dynamic perceptions and practices, so if we considered the death of Turing was not right we could move forward and express our views and be very clear and loud about our perceptions of right and wrong. In 2013, after sixty years conviction of Alan Turing (https://www.cia.gov/news-information/featured-story-archive/2015; CIA.gov, April 10, 2015), Queen Elizabeth II had granted him a formal pardon, a rare instance in the history of Britain.

Maybe one day historians would dig out the mystery behind the death of A. M. Turing, but a half-eaten apple, which was left beside his bedside intrigued huge number of people in late sixties and seventies, particularly some young, aspirant entrepreneurs, who had spent two decades in their

---

[2] Gynecomastia: It is a hormonal disease, resulting the excessive growth of mammary gland, in man due to hormonal imbalance of some sex hormones like estrogen to testosterone level.

garage and various make-shift computer labs building personal computers. They were looking for a good logo for branding of their product, and Turing's role in coding, development of modern computers, and in the field of artificial intelligence had been a huge influence on them. The sad end of Turing's life had apparently moved them in such a way, that it helped form the decision to put the graphic of an apple, with a bite taken out of it, as their product logo of the personal computers, in honor and memory of that genius, and we see the iconic logo on our popular MacBook and iPhone today. But it was Rob Janoff, the original logo maker, who mentioned it in a 2009 interview, where he said "*I'm afraid it didn't have a thing to do with it…it's a wonderful urban legend*" (Frith, October 7, 2011). Rob Janoff has further clarified that the single bite out of the apple served a practical purpose for the scaling of the object and to make sure that the specific object in the logo is not a cherry, it's an apple. Yet, it did not completely deny that the objective of using a logo of an apple, with a single bite out of it on the MacBook or Apple iPhone, had not been intended to pay the homage to Turing by the creators of Apple. The mystery of this iconic apple logo would remained there as Steve Jobs himself responded of the concern to his friend and a great actor Stephen Fry, to whether the logo was in the homage of Turing, and that made Jobs reply stoically "*God, we wish it were*" (Cooper, November 18, 2014).

If we would look at the original logo on the Apple personal computer and their accessories, which was designed by logo designer Ronald Wayne and used for the first time in 1977, it featured Isaac Newton sitting under an apple tree and a single apple hanging over his head, where it also bear a favorite quotation of Jobs from the poem of William Wordsworth "*Isaac Newton- a mind forever voyaging through strange seas of thought – alone*." But after a year, Jobs reviewed his earlier decision about Apple logo and requested Rob Janoff to create a new logo for Apple products (Biography.com Editors, April 2, 2014). Since, then it gave birth several myths and theories, why the logo of an Apple with a bite had been chosen by the founders of Apple Inc. for their unique inventions as well as the branding of it. In this concern, it needs to be mentioned here that in his biography, Steve Jobs who founded Apple Inc. with his business partner Steve Wozniak, had visited an orchard of Macintosh apples, that influenced him in the founding the name of their company "*Apple*" as he considered it "*fun spirited and not intimidating*" (Biography.com Editors, April 2, 2014).

According to the corporate gossips (as there is not enough supporting theory behind it), the iconic Apple logo represents the true reflection of Steve Jobs and Steve Wozniak's attitude to their business, in the domain of artificial intelligence and computers in the Seventies, as they were aspirant, desirous entrepreneurs taking a bite out of an metaphoric apple. Another intriguing hypothetical explanation of iconic Apple logo was Eve and Adam's first-time experience of tasting the forbidden fruit, which has been interpreted as an apple, plucked from the tree of knowledge of good and evil in the Garden of Eden, which has metaphorically been interpreted as a symbolic state of being enlightened, that a computer of a particular brand like MacBook could bestow to its users; though it has merely considered as an interpretation, where no substantiation has ever been made by Apple founders.

However, let's go for an evening-walk in the northern part of Kolkata (previously known as Calcutta), the human settlement of more than 300 years old in the erstwhile capital of British India as well as Dominique Lapierre's "The City of Joy,"[3] which once had been established by Job Charnock, the administrator of East India Company (in the pre-colonial era of British India in 17th century), in the bank of tributaries of Ganga (also known as Hooghly) in search of a young, energetic man who belonged to his virtual world and experienced a bite of an apple, for the first time in his life. In a dusk of December, 1999, in the backdrop of vibrant sounds, fragrance of hot oriental spices being cooked, the intermittent buzz of the streaming of home-going people, desperate efforts of honks of buses and yellow cabs of those who got struck in the traffic jam for a long span of time, in the northern-flank of a Upper middle-class urban settlement of Eastern Indian metropolis, Kolkata, where a young man was in front of his personal computer, which was not merely a computer, more than that, perhaps the most loyal, inanimate object in his secluded life, and it was Roop's favorite Apple MacBook. It was the first time for Roop, a city dweller of "The City of Joy," in the age of his late twenties, who was apparently not looking thrilled to take a 21 hour long flight to hop out of the City of Joy and into the City of Toronto, to start his long-cherished postdoctoral-research program at the University of Toronto, rather Roop was feeling dejected inside, looking like a person torn apart for some painful course of action. No one from his

[3] City of Joy: As Dominique Lapierre has recognized Calcutta or Kolkata in in his novel "la cite de la joie" or "The City of Joy" in 1985.

early school days or university would say that Roop was an emotional fool, rather he was recognized as a practical guy with cool brain, and little shy in his friend circle's as he hardly made any decisions that would have an impact in his practical life. Roop had gone through a stressful span of childhood after losing his parents in his early years and where he was then dumped by the relatives of his father's side to his grandparents (of maternal side) to bring him up. It was not an easy childhood and he often needed to wait for his favorite comic-book on Tintin and captain Haddock, for 2-3 months, when other kids in his setting were least bothered about it. The real-life experience in his early childhood, where emotional traits like joy or pain hardly made any difference in his way of living had moulded him to live in the world of seclusion where he hardly was touched by any emotional feeling. Anyway, from his childhood Roop brought himself up with a hard shell of chrysalis to prepare himself to achieve a high academic goal, and to go above and beyond mediocre academic dominance of his surroundings. So, it is quite unlikely to witness such a young man with realistic vision in life, to be in a state of dejection on his day of celebration of his long-cherished freedom and to come out of chrysalis to explore a new horizon. Unfortunately, Roop could not celebrate this unique moment of freedom and whenever Roop stared at the clock ticking towards 11:00pm, he was plunged in the cascading waves of sorrow, although he was apparently sitting in a social-media chatroom chatting to his friend Ira, a girl in her early thirties, from Dhaka, a good friend of his across the border of India. It was not like that they were very close or knew each other very well or for a long time either. Actually, Roop and Ira met each other during The Calcutta Book-Fair festival in 1996, the year Bangladesh was the theme country and the center attraction of the book-fair, and since then they remained good friends and Ira had visited the next two consecutive years to participate in the Calcutta book fairs, where they spent time to hanging around to explore old monuments and visiting destinations in and around Calcutta (Kolkata). This helped lay a stepping stone to help these two friends across the border to explore a bigger horizon in a human relationship and they kept in touch with each other by frequenting the social-media chat-rooms, and exchanging each other's joy and sorrow as other virtual media communities often do. After maintaining an intensive virtual friendship for three years, Roop and Ira had explored a different horizon of human relationship that made Roop want to spend last couple of hours of dusk in his small den in the Northern end of Kolkata, as his mode of self-exploration of freedom has been shattered in the shadows of a virtual dream to get Ira closer in his life. Roop's metaphoric exploration of coming out of chrysalis seems to be a roller-coaster ride of "emerging out of shadows and plunging in sorrow darkness." As the clock ticked in towards 11:00 pm, the time of his flight to North America from Calcutta airport, Roop felt, that he would not have to come back to that house of his grandparents anymore, where he had been dumped as a child, and he did not feel any emotional tie with relatives in that place. He struggled throughout his childhood and youth to get stability, as he lacked enough mental support and he felt that the people in his life hardly made any effort to understand him or his needs. For a kid without parents, it would be a difficult upbringing if nobody would take any extra effort to understand him or her, or try to compensate his mental loss by paying an extra bit of psychological support with love and affection. On the other side, he felt a severe pain that Ira, whom he considered as a good friend, appeared to be more than that in this point of his life, whom he had possibly fallen in love with, rather than just merely liked her, but he had never took any effort to ever express it, nor speak to her about it. Perhaps Roop's voice was not strong enough with low frequency (hypo-sonic) or excessively high (like hypersonic) that audible communications have failed to make its way across other side of the border to reach out to Ira, transcending through the invisible, virtual meshwork of cyber-world.

Normally, our voice box, larynx or Adam's apple, which stick out in our throat and become full grown in size and shape during adolescence of men and some women, give us voice during any sort of audible communication whether we shout, talk, sing, cry, laugh, whisper etc. and the vibrations or humming can easily be detected or felt by touching the front side of the throat with fingers. But Roop was weeping quietly in that pain, in front of the screen of his inanimate friend, the Apple MacBook, which he has to leave behind in order to fly out to his new destination, and Roop had hardly uttered his expression of love to Ira. Whilst intermittently shedding tears in front of the computer monitor during the dusk before that long flight, his chance of exploration to come out of a monotonous life and to step in an uncharted territory, made Roop feel a vacuum that might be understood and consoled by his mother, who he lost long time ago. That silent cry had left some unnoticeable vibrations in his Adam's apple which he touched and lost himself in the labyrinth of a childhood memory, where a spine from a fish got stuck in his throat and he suffered through an unpleasant experience for three days before a doctor took it out. Helpless, Roop had felt that pain of the forbidden-fruit that he consciously tried to avoid tasting, in order not to get involved with any happy friend or girlfriend, where they would then become part of his secluded life. This had now been torn apart in a bad dream of his childhood, which ultimately got stuck in his throat and he could not express it to get Ira in his life, or to be able to move on either; as it was his first love after all. We don't know whether all relations got a chance to complete a full cycle of expectations and desire like a sweet dream. In the literary world, we found a big number of stories have tend to go for a happy ending but it did not happen in reality always. In reality, it's a not necessary that all rivers transcended to an ocean but it flows on its way and human life has a metaphoric resemblance to the flowing rivers and rivulets as real-life impediments could not stop the journey of life. The matter of fact is that it's difficult to apprehend what the taste of an apple is without tasting it, as an apple appears to different entities, and in different ways to someone like Adam. If it's a matter of enlightenment, to another one like Roop, it appeared as a pink rose in the horizon of a new beginning, or to someone like Turing, it's just an exit of life, the silent kiss of sleep that helps someone to reach out to the world of peace.

However, the story of origin of the term "Adam's apple" goes all the way back to Biblical event when a devil serpent enticed Eve (the maiden woman, as par theological notion of Bible), to pluck a forbidden fruit (which has been interpreted as an apple) from "The Tree of Knowledge of Good and Evil;" and Eve took a bite and swallowed it completely (Hoffman, January, 2017; Martyris, April 30, 2017). Eve then gave the apple to Adam, (the maiden man of entire human civilization, as par the theological notion of Bible) and he also took a bite of it. Eventually, God came down around them and pronounced to stick that slice of apple in Adam's throat form a lump. According to the contemporary physiological and anatomical perspective of human bodies, the swallowing mechanism of food and making audible voice out of the voice box in our throat has resulted in the thyroid and larynx to protrude out of the front end of the throat, and has traditionally been recognized as the "Adam's apple" (Hiskey, July 13, 2010). According to the theological narrative of Genesis 2, (Hoffman, January 18, 2017; Martyris, April 30, 2017). In the Book of Genesis, 2:17, the maiden man and woman, Adam and Eve, where allowed to eat all fruits grown in the Garden of Eden, except the fruit of "The Tree of Knowledge of Good and Evil." The metaphoric interpretation of the Bible, and the eating of the forbidden fruit, which has been elucidated as an apple later, referred to as an instance of behavioral courtship of physical pleasure, which was considered to be an immoral or forbidden act. In the book of Genesis 3: 4-5, (Hoffman, January 18, 2017; Martyris, April 30, 2017) the hidden serpent of "The Tree of Knowledge of Good and Evil," enticed Eve as it said:

"And the serpent said unto the woman, ye shall not surely die: For God doth know that in the day ye eat thereof, then your eyes shall be opened, and ye shall be as gods, knowing good and evil."

Being tempted by the wisdom of the serpent, Eve has taken a bite of the forbidden fruit and shared it with Adam, and then they became aware about their nudity, and this newly acquired knowledge of good and evil manifested them to hide their nakedness by wearing the leaf clothing. As God approached Adam and Eve, He noticed them take the taste of forbidden fruit (which has been considered as sinful act) and by ignoring his sacred command they made Him angry enough to curse and send forth the serpent to remove Eve and Adam out of Garden of Eden, revoking their privileges of eternal life. So, a bite of an apple ushered cascading changes in the life of Adam and Eve as they acquired the knowledge of good and evil from the forbidden fruit, but it put them in exile, out of paradise and they lost the privileges of eternal life, instead getting a mortal life. But a reasonable elucidation of the metaphoric extrapolation of this story of the exodus of Adam and Eve out of paradise would not be considered to be shameful banishment or punishment for these Biblical characters, as God punished entities for not to abiding the God's sacred command. If we believe or agree with the theological interpretation of Abrahamic religious practices, the symbolic acquiring of the taste of forbidden fruit would evoke the self-realization and self-recognition of Adam and Eve as creator of the entire human race, our erstwhile predecessor. So, if Adam and Eve would have been considered or recognized as the creator of modern human species (either by creationists or by evolutionists), the sacred command of God not to touch the "forbidden fruit" seems to be over-protective manifest for them, our pre-historic predecessor and that would have been interpreted as emergence of another creator in success. The emergence of another creator, in presence GOD, might create a confusion about His supremacy. Hence, there is a legitimate question or concern come to the mind of creationists whether GOD were jealous about existence or tentative rise of another creator or not? Its difficult to comprehend that our crestfallen predecessor Adam might of left the garden of paradise with his soulmate Eve to pay for a bite of an Apple, and to be driven out of Eden, but it rolled out the evolutionary journey of modern human species (which supposedly started around 5-7 million years ago) where we got a chance to evolved as the most intelligent (so far) species, most sensitive species and most loving creatures on the Earth, as well who eternally inherit the sacred treasure trove of love in their heart. As an analogical successor of prehistoric Adam and Eve, modern human species have still possessed the sacred inheritance love and creation, that got infused from a symbolic bite of an apple, which helped them not only to create themselves, but made them proud as well to be part of the creations in their ambiance on this Earth, another Eden in the universe.

References

Biography.com Editors. (April 2, 2014). Steve Jobs Biography. https:// www.biography.com/business-figure/steve-jobs

Chivers, T. (October 10, 2014). Could you have a codebreaker at Bletchely Park? https://www.telegraph. co.uk/history/world-war-two/ 11151478/ Could-you-have-been-a-codebreaker-at-Bletchley-Park.html

CIA.gov (April 10, 2015).The Enigma of Alan Turing. https://www.cia. gov/news-information/featured-story-archive/2015,

Cooper, Q. (November 18, 2014). Alan Turing: Seperating the man and the myth. https://www.bbc.com/future/article/20120620-the-turing-test- of-time

Diffen.com (May 16, 2011). World War I vs. World War II. https://www. diffen.com/difference/World_War_I_vs_World_War_II

DiSalvo, D. (May 27, 2012). How Alan Turing Helped Win WWII And Was Thanked With Criminal Prosecution For Being Gay. https://www.forbes. com/ sites/daviddisalvo/2012/05/27/how-alan-turing-helped-win-wwii- and-was-thanked-with-criminal-prosecution-for-being-gay/#1603aa165cc3

Frith, H. (October 7, 2011). Unraveling the tale behind the Apple logo. https://edition.cnn.com/2011/10/06/opinion/apple-logo/index.html

Hiskey, D. (July 13, 2010). Why he Adam's Apple is called Adam's Apple? https://www.todayifoundout.com/index.php/2010/07/why-the-adams-apple- is-called-the-adams-apple/

Hoffman, D. (January 18, 2017). What was "The tree of knowledge of good and evil" for. https://knowingscripture.com/articles/what-was-the-tree-of-the- knowledge-of-good-and-evil-for

Lapierre, D. (Reissue Edition,1988). *The City of joy*. Grand Central Publishing, India. P.552.

Martyris, N. (April, 30, 2017). 'Paradise lost': How the apple became the forbidden fruit. https://www.npr.org/sections/thesalt/2017/04/30/ 526069512/paradise-lost-how-the-apple-became-the-forbidden-fruit

Muir, G. (July 24, 2019). AI- The Fifth Industrial Revolution. https:// aibusiness.com/livetiles-fifth-industrial-revolution/

Smith, M. (Reprints Edition, 1998). *Station X: The Codebreakers of Bletchley Park*. Boxtree, U.K. P.224.

Turing, A. (1950). I.-Computing machinery and Intelligence. *Mind*, LIX (236): 433-460. https://doi.org/10.1093/mind/LIX.236.433

Chapter 2

In Quest for Y-Chromosomal MRCA Adam, Mitochondrial Eve & proliferation of Humankind: In the frame of reference of genetic diversity.

Scripture [Holy Bible] identifies Adam and Eve as the sole progenitors of the entire human race, and Noah, his wife, his three sons, and their wives as the most immediate ancestors of modern humans (Gen 3:20; 9:19). Shortly after the global Flood of Noah's day, the human ancestors of the modern "races" (Mortenson,2016) or ethnic groups formed as a result of the confusion of languages at Babel (Gen 11:8–9)

---- Nathaniel T. Jeanson & Jeffrey P. Tomkens

On the 28th of October, 2018, was an interesting blog titled "Adam & Eve: Genetic Evidence," that had been written by the eminent experimental physicist Michael G. Strauss, CERN laboratory in Switzerland; Professor, University of Oklahoma, USA which had created a buzz in his reader's circle as he mentioned "*I'm not sure I can always separate the good signal from the background noise. Nevertheless in response to the wishes of my readers I will delve into the subject of Adam and Eve. In this post I'll talk about some of the scientific and genetic evidence regarding human origins and in the next posts I'll discuss some other options that may reconcile the scientific evidence with the biblical record and also examine some of the theological issues dealing with Adam and Eve.*" (Strauss, October 28, 2018). The contemporary advancement of the molecular genetics and genomics has pushed the traditional doctrine of evolution forward in a level that the entire arguments on origin and evolutionary journey of human-primate history has not only contradicted the traditional religious belief of the origin of modern human, race but it clearly formed an irreversible fissure among the creationists and evolutionists of the contemporary progressive mass of educational scholars, intellectuals, philosophers and free thinkers in the Western World, where science has found to be integral part of philosophy to a greater extent and theology to a lesser extent. In the Western World it's not only the religious preachers, priests, religious institutions and believers that influence others to follow the religious messages and practices and spread out the lesson of religious teachings, it's the intellectuals, politicians, teachers, educational scholars, doctors and even scientists (whoever believe in religious faith) that also played a cardinal role in spreading out the theological manifestos in their close circle of students, academicians, supporters, clients and followers. In the personal narrative of Professor Michael G. Strauss, one of the renowned particle physicist from USA, who has a deep-rooted faith in Christianity, the observations, mentioned above could have further been substantiated as he stated "*Many of my readers have affirmed that my writings and videos have made a strong case that God used the big bang as his method of creation and that both the record of nature and the biblical record support a 14 billion year old universe.....I have been asked to explain the scientific evidence about human origins in simple terms. ....One of my readers asked if I could comment on the genetic evidence from DNA analysis supporting an historical Adam and Eve*" (Strauss, October 28, 2018). According to the reference of traditional religious faith of Christianity, sacred scripture of in the Bible, Genesis 1-11, has narrated that Adam and Eve are the instance of spectral creation of God, which does not have any previous ancestral record and the creation of modern human on Earth has supposedly traced back to the holy emergence (Creation of God) of this maiden couple in the Garden of Eden. So from the very beginning, the traditional beliefs of the Bible, which has been strongly acknowledged by the creationists did not consider or accept the scientific principles of evolutionists, who tried to depict the origin and evolution of modern human race, had emerged out of an ancestral root-stock of primates and proto-humans in the course of biological evolution. The careful theological studies of the Bible-Book of Genesis have narrated that the creation or holy origin of Adam and Eve, the ultimate progenitor of the modern human race, had supposedly taken place around 6000 years ago. The first population bottleneck of the human population history had allegedly occurred during the First Flood and the narrative of Noah's Ark, which had supposedly taken place around 4500 years ago, this has not only been found to be overlapping, it has also been found to be corroborative to the anthropological and evolutionary history of modern human race (Jeanson, and Tomkins, September 14, 2019). The contemporary advancement in the field of molecular genetics and genomics has not only created an opportunity to study differences between the distinct molecular and genomic profiles of modern human species and other species phylogenetically related or not related to it, but it has created an opportunity to testify scientific investigations of the accountability of spiritual or theological interpretations of Adam and Eve, from the evolutionary perspective of origin of modern human species on Earth, and whether both narratives would reconcile or refute each other.

Before digging deeper let's have an overview of the basic ideas of genetics. Genetics is a specific field in biology that deals with the scientific studies of traits that are usually passed from the parents from one generation to the offspring of the next. The basic introduction to genetics introduce us to the trinity of genetics such as genes, DNA and chromosomes to understand the principle and functionalities of genetics. Let's starts with an introduction of genes. Genes are the carriers of encrypted information that determines our traits (the features or characteristics) which are usually passed on to us from our ancestors and further inherited to our successors from us. It has been estimated by scientists that there are 25-35,000 protein-encoding genes in each human cell on average (Salzberg, 2018). Actually, it has been found in every biological organism, our visible features or characteristics like skin color, hair color, color of eyes etc. we inherit from our parents. Any multicellular eukaryotic organism, specifically the human body, is made up of billions of small-sized cells, which are only visible under a microscope. Under the microscope, when we tried to specifically investigate each cell, we would be able to see one tiny, egg-shaped structure, called a nucleus, that acts like the brain of the cell. In each of the human cell there are 23 pairs or 46 individual chromosomes, half of it we got from each parent through vertical transfer via sexual fusion. Out of the 23 pairs of chromosomes, 22 pairs are autosomes, integrated into the body cells, whereas the last pair deal with the distinction of gender or determination of sex, are called sex chromosomes. Sex chromosomes are normally expressed by X and Y chromosomes, and each female normally has 2X chromosomes and their male counterpart has a singular X and Y chromosome. Under the microscopic field of each cell, the chromosomes look like small spaghetti like structures. The chromosomes in each cell are found in pairs and hundreds to thousands of genes are sequentially arrayed on each chromosomes. Normally each chromosome has been found to deal with a specific genetic task. The structural and functional units of chromosomes and genes are made of DNA or Deoxyribose Nucleic Acid. Functionally, each gene is like a YouTube video-aided recipe for a dish, instructions that offer a guide for cooking a specific dish, which would yield a unique protein (enzyme) and each protein acts like the building block of our body. Scientists have noticed that each gene on average could yield more than 10 different types of proteins, and there are around 300,000 types of proteins in our body to deal with different tasks, ranging from the formation, replenishment, servicing, function etc. of different essential tissue systems like the nervous system, respiratory systems, reproductive systems, blood and conductive systems etc. However, in order to track down our inheritance, if we introspect our genes, we witnessed that our genes briefly narrate our ancestral history. So, the confirmatory DNA investigations further determined to whether we are related to, or phylogenetically separated (long lost separation) from our siblings or not. Some DNA testing called Y chromosomal DNA testing, helps us to get to know our male ancestors, and on the other side there is another DNA, called mitochondrial DNA, which is transcended from mother to the offspring that helps to get our maternal ancestry.

In 1987, a group of population geneticists had investigated 147 mtDNA samples from women, which were collected from diversified human communities all over the world, in order to analyze and establish the genetic relationship among global communities, and then presented this in a universal chart, where a model of "molecular clock" had been adopted (on the basis of number of DNA mutations, arising in each generations) (Cann et al, 1987). Based on their research findings, Cann and his research colleagues have contended that "Mitochondrial Eve" had emerged some part of Africa around 200,000 years ago. On the other side of the story, we need to refer to the unique research work of Carlos Bustamante, an eminent population geneticist from the School of Medicine at Stanford University, California, USA, who along with a number of research associates was able to sequence the Y chromosomes of 69 males, randomly collected from diverse human communities all over the world. From this collected sample they noticed around 9000 sequence variations, which had not been found earlier (Poznik et al, 2013). The team of Bustamante had utilized these variations to create a molecular clock and contended that Adam lived somewhere in between

120,000 to 156,000 years ago. So, a comparative analysis of the men's DNA rendered the scientists to hypothesize that Eve supposedly lived between 99,000 to 148,000 years ago (Poznik et al, 2013). In another investigation of historical Adam and Eve, the unique work of Paolo Francalacci an European population geneticist from University of Sassari, Italy, needs to be referred, who along with a research team had investigated with the Y chromosomes of 1,200 men that lived in island of Sardinia (Francalacci et al, 2013). Francalacci and his research colleagues had identified 7000 Y-chromosomal variations, that were reported to be unknown earlier and they had utilized it to form a molecular clock, that helped them to map out the history of migration and the anthropological scenario of Sardinia like emergence of Neolithic populations in Sardinia, the story of historical migration of Africans as part of Roman slave trade in Sardinia etc. Most interesting part of this investigation was that it had contemplated that Adam lived around 180,000-200,000 years ago, that has found to be identical to the age of the emergence of mitochondrial Eve, who had reportedly emerged around 200,000 years ago (Francalacci et al, 2013). However, the contemporary DNA investigations and the advancements in molecular genetics and genomics have contended that the modern human race have supposedly emerged out of a single ancient mother (Biblical Eve) and ancient father (Biblical Adam) both of them have supposedly lived around 150,000-200,000 years ago, though there is a big deal of controversy in terms of the precise interpretation of variation of unique DNA sequences and the calibration of the molecular clock (Callaway, March 15, 2013; Strauss, October 28, 2018; Jeanson and Tomkins, September 14, 2019). However, this unique discovery in molecular genetics has apparently discovered the historical story of Adam and Eve, though the prehistoric existence of Adam and Eve around 150,000-200,000 years ago, which seems to be a long time ago to a number of people, as they consider the anthropological history of human species, which is not more than 10.000 years old.

As the molecular genetics of the human origin is a scientifically complex and controversial issue, there is no unanimous acceptance of such complicated issues has been found. As resolution of this it is required to address the five major issues:

1. Whether the origin and existence of the historical first couple Adam and Eve fits in the perception of genetic diversity or not?
2. Whether human lineage could have ever transcended through a

population bottleneck of two individuals?

3. Whether the origin and evolution of modern human species has any evolutionary link with ape-like primates or it has been emerged out of prehistoric human root-stock, de novo?
4. Whether the origin and evolution of modern human species has supposedly taken place around 150,000 years ago or 6,000 years ago, i.e. whether it's a prehistoric or recent event?
5. Whether the evolutionary journey of modern human race has taken place out of a micro-population or a pre-historic couple?
6. Whether the origin and evolution of modern human race has historically taken place somewhere in the continent of Africa or at the Mount Ararat?

Let's start with the genetic diversity in relation to origin and evolution of the modern human species. As par conventional perception, genetic diversity represents the gross number of genetic characteristics found in the genetic make-up of a species. Genetic diversity has found to be absolutely distinct from genetic variability, which shows the variation likelihood of the genetic characters in a species. Genetic diversity is an inherent treasure trove of any population of a species, by virtue of possession of various alleles in the different individuals of the targeted population. So, a genetically diverse population have a better opportunity of adaptation and survival in any changing environment and the matured individuals are found having better potential to produce offspring with a diverse range of alleles and surviving generation after generation, without going through the trail of genetic bottleneck (National Biological Information Infrastructure, February 25, 2011). So, the extent of genetic diversity gives an idea of qualitative index of the diversity of genes, possessed by different individuals and its hereditary relations with close or distant individuals. So, it helps to elucidate further how genetic profiles of two closely related siblings could be judged on the basis of their gene diversities (possessed by their genomes) in order to compare whether one is closely or distantly related to another individual, one the basis of similarities or dissimilarities of their genomic profile. Though there are a number of reasons that rendered the diversification of genes, molecular biologists and genomicists have contemplated that the occurrence of random mutations has resulted in the nature of genes to change, this normally occurs at the time of DNA replication (duplication) and DNA translation and is the result of a copying error. And the trick of qualitative estimation of the genetic diversity in a species, over period of time, could have apparently been judged from the rate of the gene mutation that has precisely been analyzed by the molecular geneticists and genome biologists. The molecular geneticists and genomicists have contended further that the extent of genetic diversity, which has been witnessed in the modern human population, has not been found to be compatible, if it would have been emerged out of a prehistoric couple, i.e. Adam and Eve, who reportedly existed a few hundred thousand years ago and hypothesized to be our genetic father and genetic mother respectively. Rather, the general observation on the genetic diversity of the modern human race has indicated towards the alternative hypothesis of a group of geneticists and genome biologists who had ,contended that the maiden human population on earth existed in a small population size of one to few thousand individuals, those who have existed 150,000 years ago and rendered the genetic diversity among the modern human race through vertical cascade of inheritance from ancestral generations to the descendants after completion of a number of cycles (Strauss, October 28, 2018).

Strauss, (October 28, 2018) has contended that it is apparently difficult to conclude what is the size of the prehistoric ancestral population (whether it's a couple or a population of thousand individuals) that acts as the cardinal driving force behind such genetic diversity, and the exact number of years (whether its 6000 or 200,000 years) that lead the changes of DNA configuration to create a substantial impetus of genetic diversity in the modern human race. Just referring to his own research experience as a particle physicist, Strauss has mentioned that in any experimental analysis, scientists always start with a simple experimental model and if they found that the experimental model failed to yield any results in describing the data, it requires revising the experimental protocol or model to define the data. In order to elucidate the rate of gene diversity, predicted through using mathematical models , Strauss has referred further that the accuracy level of the results has not been verified in that extent that would yield the accurate results mentioned with absolute certainty. Strauss, (October 28, 2018) has further mentioned "*Unfortunately, in genetics that feedback mechanism of making the model more complex to meet the data has not been done (and maybe cannot be done very easily). So the predictions of the models are believed even though the models are known to be idealized and simplified without independent validation. So the predictions of the models are believed even though the models are known to be idealized and simplified without independent validation. Such a model would not be accepted as valid in more rigorous mathematical scientific discipline of physics that I work in.*" In order to corroborate the inferior potential of simplistic mathematical models to decipher the complex issues like determination of genetic diversity, Strauss (October 28, 2018) has referred an instance of scientific effort of experimental introduction of a pair of Mouflon sheep from one of the Kerguelen Islands nearby the Arctic Circle in 1957. Followed by its impact of genetic diversity of Mouflon sheep in Kerguelen islands after 50 years of interval, the scientists had been surprised with the results as they were found to be four times better than the scientists had predicted using the field data in their experimental, mathematical model. So, the real-time experience has proved that undertaking of simple, mathematical model is found to be erroneous, if the experiment is related to natural science, where a number of variables are invisible or not so important to consider at the time of the experimental protocol. In order to get a clear picture of it, Strauss (October 28, 2018) has referred to the brief, exclusive narration of Dr. Fuzale Rana, who tried to make a scientific review to judge whether historic figures like Adam and Eve as a primordial progenitor of human population are justified or not, and Strauss put his own observation on the concluding remarks of Dr. Rana. In his published, peer-reviewed science journal, Dr. Rana has mentioned "*there are no published peer-reviewed articles that actually validate that the simple mathematical models are precisely valid*" and he further stated "*I was unable to find any studies that demonstrated the validity of any of the three approaches used to measure population size. Instead, I found three more studies indicating that when genetic diversity was measured for animal populations on the verge of extinction it was much greater than expected, based on the predictions derived from the mathematical models*" (Strauss, October 28, 2018).

However, like Strauss, a number of important personalities who holds their positions in the frontline in the communities of science, had reviewed the genetic evidence and corroborated further that scientific observations have promoted the traditional biblical perceptions that the contemporary human race have emerged out of this prehistoric male and female progenitors, who lived on the earth around 150,000 to 200,000 years ago. Although a big number of scientists, particularly the molecular geneticists, have contended that these two supposedly lived among a prehistoric human population of few thousands (ranging from 10,000 to 30,000 individuals), they had supposedly lived at different times. But a number of scientists like Strauss, who is a pro-Christian believer, have refuted such pro-evolutionary observations as they considered that the observation, contended by the larger group of scientists, have come-out of the so-called "simplistic mathematical models," where the analyzing potential of such model are merely too simplistic and yield results that lack scientific precision. The argument or counter-arguments of creationists versus evolutionists in favor of the emergence of Adam and Eve, and the original creation of modern human race as well as their transcendence on the Earth out of this initial, prehistoric couple around thousands and thousands years ago would failed to yield any conclusive results. When the evolutionary creationists have argued in favor of the Emergence of Adam and Eve as a creation of God, they act like the ultimate genetic father and genetic mother respectively of the modern human race, has been emerged out from the pre-human ancestors. The advocacy group members, belonging to old earth creationists, have boldly supported that Adam and Eve had emerged as a special creation of God.

The key hypothetical findings, which has partly supported by scientific analysis of DNA like "*DNA testing has shown that all humans are related to a single ancient mother and also to a single ancient father*" has been criticized along with its supplementary observations, like entire mitochondrial DNA specimens possessed by modern human race, which inherited from the maiden female individual, Eve, who supposedly lived around 200,000 + /- 50,000 years ago, has been reviewed by a number of molecular geneticists in the last two decades. As a group of molecular geneticists have further tried to draw a possible extension of such findings of their opponents by extrapolating that the distinct lines of mitochondrial DNA of all other women at the time of Eve, most likely failed to survive with the exception of Eve, that supposedly diversified into distinct f modern strands. The emergence of males in the modern human race has also been deciphered based on the possession of identical Y-chromosomes among all males, ascertained the scientific validation of hypothetical origin of males from a single male, that happened in prehistoric time. It has partially been supported by the critiques of Strauss whenever the unique discovery of Y chromosome haplogroup, in 2013 has not well known in the arena of molecular genetics yet. And it has further indicated that the emergence of "Y-chromosome Adam" in between 200,000 to 300,000 years ago has subjected to be apparently refuted by a big number of geneticists as it has not been well understood to substantiate the hypothetical observation that the modern human race have supposedly inherited all genes, transcended out of the primordial couple Adam and Eve on the earth around 150,000 to 200,000 years ago. Rather, a number of critiques of Strauss, like Martyn Cornell have contended further with firm assertions, "*we're not related to a 'single mother and father'- we're related to multiple different mothers and fathers, one each for each chromosome.*" (Strauss, October 28, 2018).

The traditional belief of creationists, based on theological narrative of the Bible, has proclaimed that all humans descended from two people or a couple, who lived around 200,000 years ago and the erstwhile human population has almost been wiped out in a pre-historic catastrophe (Biblical flood event) around 100,000 years ago seems to be pseudoscientific contention, as the scientific investigations did not have any direct evidence to provide logical support that the size of the population, rooted back to our pre-historic predecessor, that descended to the modern human race has merely been restricted to a primordial couple (Adam and Eve). Rather, the evidence from molecular genetics and genomics have indicated the size of our pre-historic, ancestral population seems to be much larger than two (Marshall, November 26, 2018). Refuting the traditional perception of the origin of the modern human race out of the primordial root-stock of Adam and Eve and the occurrence of catastrophic events related to it, Mark Stoeckle of Rockefeller University in New York, USA and David Thaler of University of Basel, Switzerland, have jointly published a research paper, published in the journal of "Human Evolution" titled "*Why should mitochondria define species*?" on the December 4, 2018, where they strongly advocated in favor of the Darwinian principle of evolution and contended further that all forms of life have originated out of a common biological root-stock and that happen in its unique journey of evolution, taking several billion years (Stoeckle and Thaler, 2018). In their evolutionary investigation, Stoeckle and Thaler (2018) have dealt with DNA barcoding, a unique effort of reading out a small part of an organism's DNA and using that information to identify it in species level. To identify an animal in species level, the scientists look for a gene called cytochrome oxidase 1 (CO1), which is not the resident of main genome, stationed in the nucleus of the cell, rather it has found to be sheltered in the powerhouse of the animal cells, called mitochondria. DNA barcoding is not considered to be a precise species identification tool like DNA fingerprinting techniques, but it has found to be an effective tool as the molecular biologists have observed that a number of individual animals (might be visually appeared little different), belonging to a certain species have identical CO1 genes in the same location, seemingly distinct from other animals belonging to another species (Marshall, November, 26, 2018). As Stoeckle and Thaler (2018) have noticed that the CO1 gene is absolute homogenous in nature in a population of any size (ranging from few hundred to few thousands) of any specific species that they considered DNA barcoding as an ideal tool that would guide them to track down the evolutionary changes of the concerned species, specifically the human species to find it out whether the erstwhile population size of the ancestors of the modern human race has been started with a small sized population, consisting of a primordial couple or with a fairly large population of few hundred individuals, and they further hypothesized that if the result of their DNA barcoding investigations revealed the nature of genetic diversity of the erstwhile populations to appear to be purged, that would indicate the very small size of ancestral population in prehistoric time, which has given rise to modern human race of approximately 7 billion on Earth. There was another challenge that Stoeckle and Thaler (2018) had to decipher, which was that the biblical catastrophe, which was reported to have taken place between 100,000 years and 200,000 years ago, would wipe out almost entire human population on Earth. This duo of scientists did not put any special importance to the supernatural occurrence of biblical catastrophe of flood, and defined further that almost each and every species would go through the bottlenecks around "every few hundred thousand years due to the rough and tumble of life" (Marshall, November 26, 2018). In an interview to Fox News, USA, David Thaler has commented "*all of animal life experiences pulses of growth and stasis or near extinction on similar time scales*," he referred the possible factors as the tentative reasons of such issues like "*ice ages and other forms of environmental change, infection, predation, competition from other species and for limited resources, and interactions among these forces*" (Marshall, November 26, 2018). Stoeckle and Thaler (2018) have contended that such bottleneck is applicable on the human species too and the intriguing comment of some molecular geneticists, engaged in genetic data of humans, in order to decipher the origin and evolution of modern human species, are found to be appeared above and beyond the rational thought of scientific observations as the presentation of the results appeared to be in a oversimplified manner that's "*consistent with extreme bottleneck of a founding pair*" (Marshall, November 26, 2018). The hypothetical idea that would like to narrate the shrinkage of erstwhile, prehistoric, ancestral population to two, after any supernatural catastrophe and the imaginary bounce-back evolutionary journey of modern human race out of that primordial couple, apparently seems to be a nice story in the genre of science fiction, though it could be treated as a big crowd puller, but it is poorly hypothesized with absolute lack of substance where major scientific issues like genetic erosion, inbreeding depression, founder effect have been left out at the sidetrack of the entire presentation.

In support of their observations, the critics of the supernatural creation of origin of the human species, the Darwinian protagonists have pointed out that the scientific observation, in the molecular level, pertaining with analysis of any gene, likely to be extracted from mitochondrial DNA, is inconclusive. As the main reason of it that mitochondrial DNA, which is considerably smaller in comparison to the nuclear DNA of any species, including human beings, has been transcended by females to the offspring so it would give the overview of the female lineage, where rest of the 50% of the genetic data on inheritance from paternal side would be remained unavailable (Marshall, November 26, 2018). The result on the basis of genetic analysis of mitochondrial DNA, often found to be insufficient to compare the genetic data, analyzed from nuclear DNA of any species. In support of this issue, molecular biologists have referred to their experience of DNA sequencing of mitochondrial DNA of Neanderthal genome and they failed to track down any interbreeding issues that happened between Neanderthals and modern human species, but that has clearly been evidenced after analyzing the DNA sequencing result of its nuclear genome. The next issue is the biblical narration of the supernatural catastrophe, which has been contemplated to occur between 100,000 – 200,000 years ago, has given a vague idea about the point of time of any accident or incident that happened in the pre-historic time as it referred to a period of time of 100,000 years, instead of a point of time as it is difficult to comprehend whether there was only one supernatural mega extinction which happened or it happened more than once over this long period of time. There was also lack of any geological evidence that could corroborate the occurrence of any biblical catastrophe in any specific point of time, that could substantiate a specific and single event of the almost extinction of ancestral human race resulting in shrinking its population size to two, either. It does not mean that the human race never felt an experience of near-extinction and repopulated further in order to ensure its place in the evolutionary race of life, as it has occurred on few occasions after large catastrophic events since the emergence of the human race on Earth. From an evolutionary perspective, it would be rather convenient to accept that a number of new species have evolved in the few hundred thousand years, in course of its complicated journey through the reticulate trail of biological evolution, where pressure of natural selection peer played the cardinal role of their evolution (Marshall, November 26, 2018). It's most cumbersome contemplation to judge the life-span of a species, as the paleontological history of a given species does not give the right clues to judge its geological age or it is often found to be missing the needed information. Technically, the perception of species varies person to person (like perception of biological species versus ecospecies to a taxonomist versus ecologists) which made it difficult to track down the dynamics of biological evolution. However, on the basis of empirical observation, the biologists have ascertained that evolutionary life span of different species varies from 500,000 years to 10 million years, before they diverged in different form or go in the way of natural extinction e.g. the evolutionary age of polar bear is around 400,000 years (Marshall, November 26, 2018).

On the basis of investigations done by Stoeckle and Thaler (2018), 90% of the extant species on Earth are assumed to be less than 200,000, and a number of scientists have construed that the age of the diverse group of extant species on Earth would be little older, as deciphering the nuclear genome would yield a more precise result than the writing deciphered in the mitochondrial DNA of the concerned species, but it has also been predicted that those would not get an indefinite period of time to sustain as factors like the cascade of swings of glacial and inter-glacial epochs in the paleoclimatic environment, that planet Earths been through for the last 2.5 million years, along with issues like overexploitation of wild germplasm by prehistoric (mostly nomadic) human beings would not render the extant species a longer lease of life. Based on their DNA barcoding experiments in quest of determining the tentative population size of our ancestors, the prehistoric human species, Stoeckle and Thaler (2018) have categorically negated the tentative existence of Adam and Eve, the founding pair of modern human race, as they have stated "*within a founding population of thousands that was stable for tens of thousands of years*." The molecular biologists have categorically accepted the shortcomings of the contemporary molecular investigations as it failed to yield a clear result on the erstwhile population size of the derived group of successors and different levels of investigations on the history of human origin and evolution have indicated and predicted about the scanty size of the prehistoric population of our ancestral lineage at the beginning of our evolutionary journey around 100,000 years ago, but that should not be confined to a couple.

For everyone, including scientists, it's not easy to step out in the uncharted territory. In 2017, the age dating investigation of the paleontological evidence, retrieved from a fossil site of Jebel Irhoud in Morocco, Northwest Africa, affirmed the age of the fossil to be around 500,000 years old (Marshall, November 26, 2018). The repeated investigations by anthropologists along with critical investigation of molecular biologists on that fossil evidence from Jebel Irhoud has found to be proved as our ancestor (Marshall, November 26, 2018). The parallel review of the molecular genetical investigation of Neanderthal (our extinct phylogenetic ally) genomes, ascertained that we, the modern human species, who have diverged out of the ancestral lineage of the prehistoric predecessor like Neanderthals, happened around 500,000 years ago, should be considered 500,000 years old, rather than engage ourselves to recognize our past, that in the labyrinth of genomic data, has apparently intended to project us under-aged modern human species (*Homo sapiens*), those who have scientifically underestimated themselves to be 200,000 years old only, whereas they have a longer evolutionary legacy on Earth.

In 2017, "*The Rise and Fall of Adam and Eve*," authored by Stephen Greenblatt, had discussed the Bible's mythical social-religious influence all over the world has not only formed Western civilization and its diversification, the famous stories of it has helped to lay down a set of socio-religious, socio-behavioral standards, as an ideal human being, with what kind of moral and ethical judgements we should have regarding the issues related to our daily way of life, like work, companionship, death, crime, punishment, marriage etc. In his literature, titled "*The Rise and Fall of Adam and Eve*" Greenblatt had narrated that the commandments of the Bible should not only be considered as a guide of socio-religious advice, or a set of moral lessons that help the human beings to be more humane by rendering them to follow a unique way of living and shaping their moral values, but he has further stated "*What's called into question is not its influence or importance, but the literal idea encapsulated within one of humanity's most famous origin stories*" (Perry, January 14, 2018).

Apparently, the critiques of theological narratives of Genesis, the opposing group of creationists as well as the evolutionists have raised a valid question to whether the ancestral couple as well as the maiden progenitor could logically populate the Earth or not? Scientifically, (in the backdrop of the scientific principles of population biology, specifically) it is most unlikely. Logically, such hypothetical scenario would have made it almost impossible for the modern human species to be on the top of all animals and being recognized as the dominant species in this age of Anthropocene in this biosphere. From the perspective of physical anthropology, the researches have engaged on determining the minimal size of the human population for a period of time, and they have presented a tentative figure in their research communications for the first time in 2013, in which it has been mentioned that around 60,000 years ago, the members, belonging to the most primitive group human population, the son of soil of Africa, have migrated out of the African continent in quest of dispersing further to ensure its widespread distribution all over the world and to be recognized as the most dominant species in the world, down the trail of its unique evolutionary journey (Perry, January 14, 2018). In their research investigations of the "Out of Africa" project, the scientists have contemplated that according to their statistical analysis, the population of around 2,250 individuals would have been moved northward to dispersed over Europe, Asia and the Middle East in order to achieve their ultimate goal to be on the top of all other species, whereas the remaining part, which has considered to be 10,000 would have been remained in Africa, around 60,000 years ago (Perry, January 14, 2018). Though such estimation is considered to be a conservative estimation of the investigations made by the population biologists, that was adopted in their statistical modelling and the possibility of an error-free result is highly unlikely, yet the population biologists have dealt with the current population size of the human populations and earnestly strived for back-tracking it in a period of interval by estimating the rate of mutations and other genetic parameters (like high level of genetic variation), to get back an estimated figure of the minimal population size of the erstwhile, prehistoric human predecessors, that were able to survive in such harsh, primitive environmental ambiance (in lack of basic amenities of life, in that time) and successfully descend to the contemporary line of genetically diversified modern human race on Earth. Both theoretical and life experience have pondered the population biologists, molecular geneticists and physical anthropologists to contend that for any successful species that struggled in the race of existence and hold the dominant position in the background of the intermittent swinging of the environmental ambiance, must have a diverse range of genes in its diverse treasure-trove of genome to cope and sustain with any tentative changes in its outer environments and to secure a long lease of life on Earth.

If we construe a hypothetical scenario, of the union of close related relatives of the human race, the children of two people are recognized as siblings and the union between them would be considered as a union between two cousins and the offspring of such close relatives would likely to give rise to populations with mental, physical, physiological and behavioral disorders. In order to avoid such a tentative appearance of physical, psychological and genetical negative impacts of developments in the future generations, incest taboo has found to be a world-wide practice across human populations, belonging to diverse socio-religious faiths and practices. It has been referred to a case-study further that the 40% of Czechoslovakian descendants, born in between 1933—1970, out of the parents, who are found to be the first degree relatives were found to be physically handicap. If we look at the family-tree history of the royals in Europe, specifically Charles II, who has been born with a number of physical and mental ailments and found to be infertile; could not walk properly till age of eight, found to have a high level of inbreeding coefficient, that he has inherited from his parents, as his parents are found to be intermarried, closely related as cousins. For keeping the royal control of the power of monarchy, a number of royals got involved in intermarriage, and their offsprings are found to be born with a number of rare genetically inherited disabilities as a result of inbreeding depression (Perry, January 14, 2018). Normally, this kind of inherited diseases are unusual and found to be less likely among human populations and the scientists have ascertained that the descendants born with such rare types of genetical disease, have usually inherited the copies of the same gene from their parents (Perry, January 14, 2018). According to the reproductive biologists and population biologists, the possession of recessive variants are not a big deal as long as we inherit it from one parent. Furthermore the molecular geneticists and genome biologists have critically observed that most of us carry a copy of lethal allele of any particular gene, the expression of it has hardly been expressed so we are physiologically and cytogenetically safe though we carry a "biologically inert bomb" in our genome (Perry, January 14, 2018). But occasional expression of the lethal effect of such recessive variants have been witnessed among some individuals, who carry double recessive variants and mostly they are found to be the descendants of the inbreeding parental lineages. If we would go through the unique, non-fiction novel of "*The Island of Colorblind*," authored by Oliver Sachs in 1998, we would have a brief narrative of the impact of natural catastrophe on human populations on an island, like a small population of the island of Pingelap of Western pacific, where all but 20 individuals of a healthy, large size human population have survived the aftermath of a severe typhoon, which happened in the 18th century. Eventually, those 20 survivors have repopulated the entire island over time and it has surprisingly been noticed by the anthropologists, medical doctors and scientists that the big chunk of population of Pingelap island, which found to be around 10%, have been suffering from a rare kind of genetical disease, caused by the presence of double recessive alleles in the gene of the patients, suffering from achromatopsia, popularly known as color-blindness. It has referred to be the unique instance of founder effect, found among the contemporary population of Pingelap, suffering from the dreadful disease of total colorblindness, triggered by the lack of genetic diversity among a human population.

In response to a hypothetical scenario, Philip Stephens from Durham University, Australia has responded that in order to avoid the negative impact of "Founder Effect":

a. 50 individuals would be enough to keep continuing the journey of human population.

b. 500 individuals would be enough to form a unique gene-pool that all offsprings would be able to adapt in a new environment and

c. 500-5000 individuals would be able to form a genepool that would be able to sustain the random genetic erosion, as the replenishment would be uninterruptedly transcended from one generation to next generation here.

The scientific observations made by the genome biologists have compared the genetic diversity of the modern human species (*Homo sapiens*) to its phylogenetic close ally, the chimpanzee, and critically observed that the overall genetic diversity of the Chimpanzee is supposedly higher than that of modern human species, as they have contemplated that the modern human species went through a genetic bottle-neck between 50,000-100,000 years ago causing death of nearly the entire human population, just leaving 1000 only individuals alive and we, the 7 billion population of human on earth have supposedly descended from them (Perry, January 14, 2018). From the perspective of modern advancement of science, scientists have contended that we have supposedly descended out of genetic Adam and mitochondrial Eve, though traditional existence of Adam and Eve (according to the belief of creationists), has yet to be testified. The scientific investigations on the genetic Adam has elucidated further that the "Y-chromosome," the signature chromosome, possessed by the genetic Adam, has initiated to be transcended from him to the following generations, whoever evolved in Africa between 125,000 to 156,000 years ago and intermittently kept on passing it till today (Perry, January 14, 2018). Though a number of evolutionary biologists did not agree with the specific pre-historic time slot, but there is no doubt the mode of inheritance of Y-chromosome from genetic Adam to the contemporary descendants. The evolutionary biologists have firmly expressed their observation that "Genetic Adam" might not be contended as the first male on Earth but he is the one who passed on the genetic information to the descendants, specifically to the contemporary extant males, who have traversed through a long trail of biological evolution to exist in the current regime of Anthropocene. On the other hand, there is a unanimous scientific observation about the existence of the "Mitochondrial Eve" as the first female, that she kept passing on her mitochondrial DNA across the females of the entire human populations, evolved for thousands and thousands of years (Perry, January 14, 2018). The scientists are not sure whether the Genetic Adam and the Mitochondrial Eve had known or met each other (as a number of investigations substantiated that emergence of Genetic Adam and Mitochondrial Eve happened in the different prehistoric period of time slots) in the pre-historic period of time or whether they should be considered as the primordial couple on the planet Earth or not, but the scientists are absolutely sure about the passing on of their genes to the extant human lineage since the initiation of the human history till today.

References

Callaway, E. (March 15, 2013). Famous “HeLa” Human Cell Line Gets Its DNA Sequenced.
https://www.scientificamerican.com/article/famous-hela-human-cell-line-gets-its-dna-sequenced/

Cann, R., Stoneking, M., and Wilson, A. (1987). Mitochondrial DNA and human evolution. *Nature* 325: 31–36. https://doi.org/10.1038/325031a0

Francalacci, P., Morelli, L., Angius, A., Berutti, R., Reinier, F., Atzeni, R., Pilu, R., Busonero, F., Maschio, A., Zara, I., Sanna, D., Useli, A., Urru, M. F., Marcelli, M., Cusano, R., Oppo, M., Zoledziewska, M., Pitzalis, M., Deidda, F., Porcu, E., … Cucca, F. (2013). Low-pass DNA sequencing of 1200 Sardinians reconstructs European Y-chromosome phylogeny. *Science*, 341(6145): 565–569. https://doi.org/10. 1126/science.1237947

Greenblatt, S. (2017). *The Rise and Fall of Adam and Eve. W.W. Norton & Company.* P.432.

Jeanson, N.T., and Tomkins, P. (September 14, 2019). Genetics Confirms the recent, supernatural creation of Adam and Eve. Chapter10.
https:// www. answersingenesis.org/bible-characters/adam-and-eve/genetics-confirms- recent-supernatural/-creation-adam-and-eve/

Marshall, M. (November 26, 2018). No, Humans are probably not all descended from a single couple who lived 200,000 years ago.
https://www.forbes.com/ sites / michaelmarshalleurope /2018/11/26/no-humans-are-probably-not-all-descended-from-a-single-couple-who-lived-200000-years-ago/#30e8441f7cd8

Mortenson, T. (2016). *Searching for Adam: Genesis & the truth about man's origin.* Master Books. P.528.

National Biological Information Infrastructure (February 25, 2011). Introduction to Genetic Diversity. U.S. Geological Survey.
http://www.nbii.gov/portal/server.pt?open=512&objID=405&PageID=0&cached=true&mode=2 &userID=2

Perry, P. (January 14, 2018). Does the story of Adam & Eve work scientifically?
https://bigthink.com/philip-perry/does-the-story-of-adam-eve-work-scientifically

Poznik, G. D., Henn, B. M., Yee, M. C., Sliwerska, E., Euskirchen, G. M., Lin, A. A., Snyder, M., Quintana-Murci, L., Kidd, J. M., Underhill, P. A., and Bustamante, C. D. (2013). Sequencing Y chromosomes resolves discrepancy in time to common ancestor of males versus females. *Science* 341(6145): 562–565. https://doi.org/10.1126/science.1237619

Sacks, O. (1998). *The Island of Colorblind.* Vintage. P. 336.

Salzberg S. L. (2018). Open questions: How many genes do we have?. *BMC biology*, 16(1): 94. https://doi.org/10.1186/s12915-018-0564-x

Stoeckle, M.Y. and Thaler, D. S. (2018). Why should mitochondria define species? *Human Evolution* 33(1-2): 1-30.

Strauss, M. G. (October 28, 2018). Adam and Eve:Genetic evidence.https://www.Michaelstrauss.Com/2018/10/adam-and-eve-genetic-evidence.html

Chapter 3

Origin of Modern Human Species in the Evolutionary Labyrinth of Adam, Eve & Apes: I (Traversing through Palaeontological Observation)

There is a popular image of human evolution that you'll find all over the place....On the left of the picture, there's an ape....On the right, a man ...Between the two is a succession of figures that become ever more like humans....Our progress from ape to human looks so smooth, so tidy. It's such a beguiling image that even the experts are loath to let it go. But it is an illusion.

---- ***Bernard Wood***

Every object has been attributed with more than one perspective where the precise narration of such an object depends on the discretionary visualization potential of the narrator, likewise, the completeness of the fossil record is such an issue, which gives distinctive narratives to its narrators. When the creationists have contended with unique ideas and interpretations about the formation, existence, and availability of fossils and its interpretation of it in defining the appearance of living beings (ranging from small creatures and serpents to modern human beings) according the Biblical narratives (according the Abrahamic religious faith and practices), under supernatural power of special creation, the evolutionists have intended to put together the state of art of biological and anthropological researches to accommodate the contemporary advancement of molecular genetics and genomics, virology and molecular evolution supplemented to the conventional practices of palaeontological investigations to narrate the origin and evolution of life, that descended from a prehistoric, simple form to the diversified, complex, derived one nowadays. Fossilization is recognized as a scientific and amalgamated mechanism (geological process to a greater extent and biological process to a lesser extent), that any plant or animal body goes through in the state of rapid burial that the fine geological particles of sedimentary rocks preserve it from further decay or hold its replica or complementary template that persisted on the earth through the ages with substantial evidence (either complete or partial) of the pre-historic existence. Let's find out whether there is any direct relation of fossil evidence as well as the fossilization process, to the biblical perception of special creation. From the outline of the brief depiction of fossilization, mentioned above, it seems that it requires a congenial environment (geological abode) for the successful occurrence of fossilization that hardly is rendered by the terrestrial environment. This kind of empirical observation has been made based on the availability of most fossils formed in the ambiance of water-mingled sediments, that might be exposed to submerged or emerged state and have gone through the process of fossilization.

The geologists have observed that around 75% of sedimentary rock, that has been found in the fossiliferous specimens, has also been found in the stratified rocky crevices, found all over the landmass of different continental shelfs of Earth (Peppe and Deino, 2013; Menton, May 29, 2018). According to the Biblical narrative of Noah's day (of 2348 BC), as the judgment of God on man's wickedness, that there was a year-long flood, that had destroyed the pre-flood by burying down billions and billions of creatures, reshaping the continents with multiple stratifications of rocky substratum and God had spared only eight righteous people along with representative of all land animals on the Ark of Noah to render them a chance of re-flourishing at the end of the global catastrophe or post-flood regime on Earth (Millman et al, December 5, 2012). So, from Abrahamic religious perspective, the historical Flood of Noah's day would have rendered favorable conditions for deposition of sediments and supposedly created an ambiance in that prehistoric regime to surge an instance of fossilization on the global level. So, whenever, this kind of substantiating elucidation helped the supporters of special creation to promote the ideas of the fossils as the result of Global flood, around 4000 years ago, the evolutionists are found to be reluctant to accept that a prolonged exposure to flood could render a surge of fossils (Hendel, January 08, 2021). But the critical investigations of paleontologists have revealed that most of the terrestrial animals were found to be exposed to water bodies, to be transformed or go through the process of fossilization. The scientific investigations have further ensured that it's the prolonged duration of flooding, rather than local flood or upsurge of water bodies adjacent to coastal regions which could trigger the process of fossilization. However, the creationists have earnestly believed that living beings (plants and animals) have hardly changed to any extent and most of the living plants and animals, existed in the pre-flood regime, had become extinct. So, according to the perception of creationists, the features of fossil specimens, supposedly identical to the extant, diverse array of animals and plants (belonging to families, genera, and species) on Earth (Menton, May 29, 2018). So, according to the supporters of special creation, the availability or existence of the transitional, intermediate or progressively evolved fossil specimens are unlikely to be found (either genus or species level) or existed among distinct kind of living beings (animals or plants), which inevitably contradicts the perception of evolutionists. Most of the evolutionists have contemplated that enormous species have evolved in last 600 million years history of biological evolution, and a small number of it had been able to survive in the struggle of existence as it followed the Darwinian doctrine of "Survival of the fittest," and 2 % of the species on Earth were preserved in fossilized form (Hayden, February, 2009). Even there would have been millions of obstacles to investigate the evolutionary history of any particular living being on Earth, the availability of an transitional form of fossils or evolutionary link between two distinct form of living entities seems to be like winning the jackpot as it helped to narrate its dynamics as well as the progress (or regress, as evolution does not have any particular direction of movements like simple form to complex or vice versa) of evolution, and to render the scientists to track down its evolutionary trail from ancestral, prehistoric or extinct form to the derived, recent or extant form of inheritance.

Based on the availability of a diverse group of fossil specimens, represented by diverse group of living organisms, belonging to distinct habitats, the most number of fossil specimens were represented by marine animals. It has been found there are around 250 million fossils found to be cataloged in the world-wide catalog of fossil specimen, and 250,000 species (Wilson and Peter, 1988; Menton, May 29, 2018). Scientists found furthermore, if 95% of these fossil specimens belonged to marine invertebrates, the remaining part has been represented by the marine algae and plants. The overall findings of fossil specimens ascertained that 0.25% of the animals are vertebrates and most of the fossilized specimens belong to

Pisces (Menton, May 29, 2018). Paleontologists have observed that out of available vertebrate fossil specimens, only 0.001% of fossil specimens belonged to hominoids and hominids, represented by monkeys, lemurs, chimpanzees, gorilla and modern human species (Rocatti and Perez, 2019). The general trend shows that mostly slow-moving marine invertebrates (having hard shells) or the living beings belonging to marine habitats have been favored and preserved in this process of fossilization, rather than their terrestrial counterparts. So, it appeared that relative completeness of the fossil record of some group of animals like terrestrial vertebrates (as they have very little number of fossil representatives in respect to their extant representatives) are not providing any feasible figure form that the scientists could determine the extent of the terrestrial vertebrates found to be widespread in distribution all-over terrestrial biomes in the world. But this is not the real scenario if we look at the analysis, made on fossil specimens of the terrestrial vertebrates, reaching its distinct taxonomic level. The paleontologists have elucidated it further that of the total representative of contemporary terrestrial vertebrates, belonging to 43 orders, one or more fossil representatives have found to be belonging to 42 orders and that represents 97% of the extant vertebrates descended out of ancestral root-stock and if we look at the family level, 329 families of terrestrial vertebrates used to have 261 prehistoric ancestors which represents around 79% representation (Denton, 2002). In 1968, in his meticulous publication, entitled "*Pleistocene Mammals of Europe*," author Bjorn Kurten has mentioned that his studies revealed that around 88% of living species of European mammals have been represented by the presence of its fossil representatives and this figure has gone up to 99% level for extant mammals, distributed over globally. As long as the age of origin and evolution of vertebrates are concerned, they are considered to be evolved recently compared to the invertebrates. So the relative completeness of fossil records would not seem to be biased to glorify the evolutionary history of the marine invertebrates that is related to the Biblical narrative of Flood hypothesis. The evolutionists have come up with a new perspective as they contended that the relative completeness of the fossil record has conveniently integrated to the richness of the diversity of the vertebrate fossils, those supposedly evolved slowly and have gone through a series of transitional nodes of evolution and representing a series of transitional forms between two distinct forms, that helped to explain the community of science which the divergence of any derived form of life is a slow but continuous evolutionary ongoing process that adapted to the intermittent evolutionary changes and remained alive in its derived form, after being descended out of ancestral lineage.

The availability of enough primate fossils rendered our formal learning process restricted to have a direct exposure to gather hands-on experience from studying it rather than relying on studying its casts and studying the secondary data published in different scientific literature. A brief enumeration of fossil specimens (including a distinct form of bones or teeth) on distinct form of fossil remnants of human species ascertained there are around 6000 distinct types of human fossils cataloged by Smithsonian National Museum of Natural History, collected from different fossil sites all over the world (March 15, 2010, http://humanorigins.si.edu/evidence/human-fossils). The scientists have noticed that in the world-wide collection of fossils of primates, most of the specimens belonged to monkeys and apes, with a considerably lesser number belonging to modern human species. In most of the cases, either in blogs on evolution related discussion or in the scientific literature related to physical anthropology, the phrase "the fossil tells us" is randomly used to elucidate any interpretation of the author (Menton, May 29, 2018). Actually, "fossils" don't tell us anything about past it's the analysis and interpretation by scientists (paleontologist) that deciphered the past of that erstwhile species, found to be disappeared a long time ago or in the prehistoric state in respect to the contemporary backdrop of space and time. The elucidation of the past of any fossil is considered to rely upon their "starting assumptions" and that "starting assumptions" of any evolutionist, to get an idea of the origin of modern human species, has remarkably different from the interpretation and its elucidation made by a supporter of special creationism. Let's make a brief review of the interpretation and assumption of fossil specimens of primates in general and human species in particular to have an idea on the origin of modern human species on Earth.

A. Interpretation and assumption of the supporter of special creationism about the origin of modern human species from fossil evidence:

Most of the followers of Abrahamic religious faith do believe that God created this universe and man and all other living beings in distinct forms (comprised of a diverse range of animals and plants) as part of this universe by executing discreet wishes of His sacred Spirit and Holy command. So whenever the supporters of principle of special creationism came across the intensive studies of primate fossils, their analysis had come up with straight-forward interpretation that recognized the natural process of extinction of any living being and its existence as distinct entity (with an ignorable trace of variations) but they never agreed upon that the existence of transitional form of variation between two distinct forms supposed to be recognized as part of the intermittent process of speciation, as the principle of special creationism did not recognize species (according to the model of taxonomic hierarchy), rather they recognized the living being, recognized by its family (Menton, May 29, 2018). So, the possibility of the emergence of a new kind of form or species out of the transitional node of ancestral lineage had never been acknowledged among the followers of special creation, belonging to the Abrahamic religious followers. So, based on the interpretation of paleontological evidence, the believers of special creationism strongly believe that its God, who created Apes and man (modern human species) as distinct forms of entities on Earth but He would never supposedly create any transitional living being like Ape-man and the consideration of any possibility like divergence of man (modern human species) out of Apes, was considered to be the interpretation of God's creation against his holy prudence. Though a group of moderate religious followers has tried to accommodate the evolution of the principles as part of God's ultimate will as they tried to present "religion" itself as part of evolution, but they failed to define what was the cardinal role that God played in creation and evolution of the universe and its living entities on Earth. If we would go through the historical narratives of the formation of the universe through the cascading nodes of cosmic evolution, inorganic evolution, organic evolution and biological evolution, leading to the present state of life on Earth, a long historical journey that had been initiated around 14 billion years ago (Menton, May 29, 2018; Kundu, 2018). Logically, the evolutionary journey of "religion" seems to be co-evolved with evolution of the modern human race on Earth, approximately 4 to 5 thousand years old, so it was unlikely to judge, analyze and interpret the entire historical journey of the formation of the universe and living organisms, mostly obscured, through a short-stretched, symbolic view-finder of time. Dr. David Menton has made rhetoric comment in his review on the contemporary literature: "Searching for Adam," authored by Terry Mortenson as he said: "*Theologists and Christian laymen should be aware that evolution is a jealous 'god' that will have no other gods before it*" (Menton, May 29, 2018).

B. Interpretation and assumption of evolutionists about the origin of modern human species from fossil evidence:

The evolutionists assumptions of evolution intrigued them to come up with a scientific and materialistic explanation which is considered to be the origin of natural material in this universe, where evolutionists don't see any scope of divine influence to come up with any intellectual design. Most of the time, evolutionists argued in favor of any creation whether it has any naturalistic (materialistic) identity or perception, so if it comes to the origin and evolution of the modern human species or man it is convincible but when it is construed that "religion" has been emerged out of evolution, the perception of evolution is found to be somewhere in "catch 22." Often, the extrapolation of materialistic interpretation of evolution took its interpretation too an extreme level that David Menton has referred an incident where an evolutionist of Humboldt University, Berlin observed chimpanzees throwing rocks at the trees, that has further been interpreted as worship ritual of Chimpanzees and tried to interpret it as evolution of religion among apes, and presented by the concerned evolutionist in the heading of "Chimps believe in God!" which sounds like an exotic idea trying to stretch out evolution beyond its possibility of scientific extrapolation (Bushfield, March 4, 2016). The principles of evolution has supported the origin of human from nonhuman rootstock, but contemplating such to extend the hypothesis that a one-celled aquatic cellular organism as the ancestor of modern human species sounds to be speculative, when its answered by any paleoanthropologist but its never been asked to them whether human species had diverged out of the ancestral lineage of apes or not, as paleoanthropologists are overqualified for facing this question as based on their investigation of fossil specimens, they could figure out which type of ape would have probably been considered to be the prehistoric ancestor of human species. The foundational assumptions of evolution based on the evolutionary relatedness or proximity between two entities of similar taxonomic rank based on the extent of identical structural, functional or genetical characteristics between these two living entities but such comparison evolutionary relatedness has considered to be apparent when evolutionary relatedness is determined between two species. If evolutionary relatedness is determined between modern human species, apes and prehistoric dinosaurs are concerned, apes are considered to be evolutionary closely related to modern human species, but contemporary phylogenetic investigations have revealed that the prehistoric dinosaurs, even the extinct amphibians like *Acanthostega* and Tiktaalik, those evolved around 398-330 mya, have also been considered to be distantly related to the human species in course of our self-exploratory introspection to go through the evolutionary history of vertebrates on Earth (Riley, December 30, 2008). In 1955, in "What is Science?" by Julien Huxley, an eminent evolutionist of his time, has shared his eloquent perception on the functionality of biological evolution as he said: "*Evolution is the whole reality, a single process of self-transformation.*" On the other side, there is a possibility that identical structural and functional features are found in distant phylogenetically related species or even phylogenetically unrelated species (not monophyletic species), which rendered the concept of evolutionary relatedness weaker. But it is quite interesting to note that a side-track principle of evolution, which is recognized as "convergent evolution" has accommodated such phylogenetically unrelated species those contemplated to be independently evolved with identical traits to adapt to similar habitats or sharing identical niches under the broad-band network of evolution. A typical example of convergent evolution is the evolution of wings or flight organs for volant adaptation, found among the distinct type of animals like flying insects, bats, birds and pterosaurs. There is a good analogical reference related to the convergent evolution of modern human species is the common cultural adaptation in the backdrop of common environmental conditions, maintained by the people belonging to different ethnic groups, descended from different ancestral cultural lineages. From a realistic viewpoint, evolution could accommodate a series of hypothetical narratives, related to creation or emergence as long as it did stay in the arena of a naturalistic (materialistic) periphery of science without any compromise of promoting ideas that creation of any biological entities are related to intellectual design or divine creation.

To any paleo-anthropologist, the narration of human evolution means dealing with a comparative analysis of anatomical and structural features of primates mainly the ancestral lineage of apes, contemplated as the successor of modern human species and the diverse type of modern human species, occupied in diverse ecological niches. For comparative anatomical investigation, the paleo-anthropological studies of the extant specimens of the modern human species had revealed a partial picture depicting the evolutionary history of the human race and without undertaking palaeontological studies, i.e. the critical investigation of fossil specimens of the pre-historic apes remains an impossible endeavor. It requires the expertise on the hard tissue anatomy, that primarily deals with palaeontological investigations of skeletons, bones and teeth (as the preservation of soft tissue of primates is most unlikely) of primates to distinguish the phylogenetically recognized ancestral members of apes from their descendants of the modern human species. To tender a feasible comparative account of anatomical characteristics of the hard tissues of fossil specimens of primates, paleo-anthropologists, paleontologists and evolutionary biologists have primarily focussed on a brief comparative account of the features that helped to distinguish apes from their descendants and it has helped narrate the evolutionary progress of the modern human species, and these key-features are as follows (Hunt, 2016):

1. Size of the skull and comparison of small size brain vs. large size: From a paleo-anthropological perspective, size of the skull of the primate fossils is considered to be a key feature to assume the volume or size of the brain of distinct categories of primates that could help the scientists to distinguish the modern human species from their phylogenetic lineage of predecessors. The critical anatomical investigation of apes revealed that the carinal cavity of the Hominins are in the ranges of 300-500 cm^3, whereas the carinal capacity of modern human species is found to be little bigger than apes, ranging between 700-2000 cm^3, which is around 3 times larger than apes. But the higher level of intelligence and cognitive potential in humans does not solely depend on the carinal cavity of the brain as the structural formation of nerve cells and its connectivity, the passage of nerve impulse, etc. that would have not reflected in the fossil of human skulls.

2. Little convex or flat shaped forehead vs. curved forehead of skull: From the side view, humans have a horizontally curved upper face and forehead having visible eye sockets or orbits whereas the base of the eye sockets is found to be flat in apes and from the skulls side view, the eye sockets of apes are unmasked by its relatively flat in comparison to Sapiens, which is found to be stretched upwards.

3. Sideview of the face of the skull, vertical vs. slope articulation: The scientists have critically observed that the simple anatomical investigations like having side-view of the fossiliferous skull of primates helped them to distinguish the modern human species from rest of the apes. From side-view of the skull, human skull is looking like vertically aligned whereas it is found to be sloped (where the imaginary line of slope seems to appear like an inclined walking stick, with a little curved handle), extended from the eye sockets and stretched up to the upper jaw and technically it has been expressed as the angular position of the face in respect to zygomatic arch (zygomatic arch is the bridge of bones, the temporal bone grew at the side of the head around upper jawbone and it also includes zygomatic bone at cheek). The human face is vertically straight, perpendicular to the zygomatic arch, whereas, in apes, it is obliquely angular, sloped or sliding in nature.

4. The position of nasal bones - protruding vs. flat in shape: Anatomically, a pair of nasal bones found to be met at the upper-end of the nasal cavity of the primate skull. In the human species, it protrudes out to form a nose bridge, on which the eyeglass could rest on, whereas in case of apes, it has found to be flat. Along with it, a nasal spine has also been found to form a septum and separate two nasal tubes, that are found to be lacking in nasal architecture.

5. The distinction of enamel and teeth in primates: As the teeth and jaws of the primate fossil are calcified hard part of the tissues, it has been found plenty from different fossil sites. The incisor and canine teeth of humans are found to be relatively smaller than molar, in comparison to its ancestral lineage of apes. Enamels are found to be one of the hardest and important biological material of the teeth of primates that would help to distinguish man (those have thick enamel in their teeth) from apes (those have considerably thinner enamel coating in their teeth). Though exceptionally thick enamel coating has been found in the teeth of some apes like orangutans.

6. The distinctive function of gluteus medius muscle in bipedal vs. quadrupedal locomotion: The investigation on locomotion of extant and fossil apes, particularly the locomotory organs, elucidated that apes walked in quadrupedal fashion (with two hands and two legs), with several exceptions like pygmy chimpanzee that is used to in 25% of bipedal locomotion whereas the modern human species could walk in bipedal fashion (with the help of two legs). It has been noticed by the anthropologists that the pelvis or hipbones are stabilized by the muscles of gluteus medius (that move side-wise of the hip) keep the right hip parallel to the ground, during locomotory movement (bipedally) of

humans but it rendered the upper part of the body in swings from one side to another side (popularly called "ape walk" as the gluteus medius of apes act as an extensor make the posture of the body leaning front-ward, during its effort to walk bipedally.

7. Carrying angle of knee and locking knee: The evolutionary biologists and paleo-anthropologists have contemplated that the meeting point of the leg bones like femur and tibia at the knee, femur bone form a carrying angle (also called valgus) with tibia which is around 9 degrees, that helped human to walk bipedally, whereas it has found to be 0 degrees for other apes, engaged in quadrupedal locomotion. It has also been observed by the scientists that for human species, the human knee locks when it straightened up and unlocked when it's bent down so it rendered the human beings to stand on it feet straight and need to bend little at the time of walking but apes could not stand straight-way and walk-in little bending posture as its knee never locked.

8. Feet bones - alignment of big toes for distinctive locomotion: The formation of human feet is remarkably different from the feet of apes. The big toes of human are found to be almost straight and pointed forward and maintain alignments with other toes of the feet rendering it fit for walking on flat surfaces, whereas the big toes of apes aligned inward found to be arranged in angular direction, rendered the apes fit for walking and grasping, to climb on the trees.

9. Hand Bones - distinctive structure of thumbs: The meticulous studies of the hand bones of primates revealed that the thumbs of humans is found to be longer than other apes and it has found to be extended beyond the knuckles of forefinger, whereas the thumb of apes hardly reach up to the knuckles of forefinger of apes. The other fingers (except thumb) in apes are found to be much longer and curved than humans, helping the apes to climb on trees, and that rendered the apes to be recognized as "knuckle walkers."

The paleoanthropologists have mentioned about the possibility of several errors that might render the scientists to make an erroneous analysis unless profound knowledge and expertise would be there during studying the primate fossils like the size of the skull of the primate in younger age found to be much larger than the adult members of same species, mainly due to high growth of nerve tissues of primates in youth so if a proper idea on the developmental cycle of primates are not there, someone could interpret with the fossil specimen with large body size and bigger skull size as an ape-man close to human species and which is erroneous interpretation of the fossil specimen indeed. The scientists observed that sexual dimorphism is a big issue that might misguide a scientist to consider a fossil of same species with larger skull size as different species, but the matter of fact is that the male member of any primate-like male gorilla is found to be considerably bigger than its female counterpart. If the scientists did not have any proper knowledge about osteo-degenerative disease, that occurred in the Neanderthal communities in prehistoric time, fossil expert like Marcellin Boule did not consider the anatomical features of the Neanderthal fossil, known as La Chapelleaus-Saints, in 1911, suffered from such pathological disease and failed to recognize it as the phylogenetic ancestor of modern human species, until Straus and Cave re-examined the fossil in 1957 and let it free from evolutionary bias (Trinkaus,1985; Menton, May 29, 2018).

Most often the scientists use to come across fragmented fossils rather than the entire skeletons been found in articulated form. For primate fossil reconstruction, the paleoanthropologists require a thorough knowledge of primate anatomy. A unique example on the reconstruction of the skull of KNM-ER 1470 specimen needs to be considered here when scientists have put a meticulous effort to reconstruct 150 pieces of skull fragments and finally recognized it as the skull of *Homo habilis* or *Homo rudolfensis* (Menton, May 29, 2018). But it created confusion as the volume of the carinal cavity has been measured around 800 cm3 and the side-view of it affirmed its sloping appearance of the face, rendered some paleoanthropologists to consider it as a skull of apes, rather than hominoid skull (Menton, May 29, 2018). Though careful investigation of primate fossils have helped the scientists to recognize and categorize a diverse kind of fossils in human or ape categories, but paleoanthropologists have tried to define the prehistoric emergence of ape-men by traversing through the evolutionary trail of three hypotheses, which have been discussed herewith (Menton, February 25, 2010; Radford, April 25, 2011):

A. Upgrading fossil apes to ape-men: To find out the emergence of an ape-man, paleoanthropologists have tried to unearth the fossil records of apes to promote depicting the emergence of human-like entities out of apes like ancestral lineage. As the extant apes and humans share several features, mainly the skeletal structural features, scientists have strived to find-out identical features between fossil apes and humans, which might not be treated as evolutionary evidence. Now the main concern is whether the fossil apes are different from modern apes or are more identical to modern humans? Evolutionists have contemplated that out of four genera of hominins, those are considered to be ancestral lineages of humans, three prehistoric representatives: *Ardipithecus*, *Australopithecus* and *Paranthropithecus* are looking like apes. The structural and behavioral features of these members of hominins found to be minimally distinct from the modern humans (those contain a series of advance features) belonging to genus Homo so that leaves a visible gap down the evolutionary trail from apes to humans through the transitional state of ape-man as the missing-link.

B. Downgrading fossil humans to ape-men: The paleoanthropologists have contemplated that the existence of pre-historic human fossils should be strategically recognized as ape-man irrespective of in-depth anatomical investigation of its hard tissues. In this regard, the scientists have referred the prehistoric human-like footprints, discovered in the fossil site of Laetoli in Tanzania and it has further been considered that the foot-print has been set by the ape-like ancestor of humans, recognized as *Australopithecus afarensis*, that has reportedly evolved 3.66 mya. Though Australopithecus and modern humans possess some anatomically identical characteristics, the pre-historic emergence of it left an evolutionary pause to be considered an early representative of the modern human species and that left the possibility of ape-man, as an evolutionary transition in the history of human evolution on Earth.

C. Composite formation of Ape-Men (by put together fossil bones of apes and fossil bones humans): One of the most popular examples of ape-man is considered to be a combination of bones of apes and the modern human species, where the fossil remnants, collected from the fossil site of Piltdown, England and the reconstruction made by Charles Dawson, the professional medical doctor and amateur fossil hunter, who recognized it as "Piltdown man" in 1912 and it has been stored in Natural History Museum, London, U.K. The critical paleontological investigation recognized its jawbone matches apes, whereas its teeth and skull found to be identical to the modern human species, so it has been recognized as "Dawn man" also. The age determination investigation ascertained its age found to be between 500,000-1,000,000 years old and contemplated to be the distant relative of the modern human species or prehistoric apes and an well-written feature had been drafted by W.P. Pycraft, published in the Illustrated London News, on December 28, 1912 (Agin, 2006). In 1953, a critical investigation of this composite fossilised ape-man has unfurled its pseudo-scientific identity as the skull part of it has been recognized as a 500 year old human skull, with a canine tooth found to be collected from a chimpanzee and the jawbone belonging to a female orangutan. The scientists recognized it as "intentional deception" or "unintentional mistake" (Menton, February 25, 2010; May 29, 2018). The primate fossils are found to be scattered in the fossil sites, which made it often difficult to the scientists to judge whether those fossilized hard tissues belonged to one or more primates. But most often, the true identity of the discovery of the composite fossil-like "Piltdown Man" has ended up in the further exposition of a scientific hoax, like the emergence of "Ape-Man," down the hypothetical trail of evolution.

If we look at the old system of taxonomic classification of plant and animals, the grouping of certain species or genera are found to be designed based on the anatomical similarities and relatedness, but the contemporary effort of the taxonomists to include the identical features on embryological, genetic and molecular genetics and evolutionary relationships rendered it complicated. The modern trend of systematic classification of animals including humans and its ancestors, based on identical molecular data and evolutionary relatedness, rendered the works of molecular phyllogeneticists in the frontier of science and the contemporary findings of molecular phylogenetics put the previous models of taxonomic classification obsolete and evolutionary relatedness has appeared to be the only criterion to be considered with priority. Another issue, that rendered the taxonomic classification chaotic is the dynamic nature of nomenclature of a certain taxonomic rank as the taxonomic judgment of scientists to draw a new taxonomic affinity, merging of two distinct categories of animals or plants of distinct taxonomic ranks into same or splitting off an existing one to establish it in a distinct category is found to be never-ending process. So, when systematic zoologists, paleoanthropologists and molecular phyllogeneticists have endeavored in the taxonomic classification of humans and their fossil ancestors, and have no other choice but to go through the chaotic kaleidoscope of evolutionary history of human-related to its origin, evolution, and diversification. To get through the gateway of the evolutionary history of humans, we came across the most popular term, called "Hominid," which refers to modern human species, modern and extinct Great Apes and their predecessors. Recently, a new term "Hominin" has found to be popular in defining the evolutionary history of modern human species, where modern human species, that belonging to the genus

"Homo" and its tentative immediate ancestors: the genera of *Ardipithecus*, *Australopithecus* and *Paranthropus* have been taken into account by the molecular taxonomists. In September 8, 2015, Nathan M. Young, the eminent scientist in Orthopaedic Surgery, University of California along with three research colleagues in human evolutionary biology Terence D. Capellini, Neil T. Roch, and Zeresnay Alemseged have published their research works, entitled "*Fossil hominin shoulders support an African ape-like common ancestor of humans and chimpanzees*" in the Proceedings of National Academy of Science, USA, and this rendered a possibility of coordinating the scientific investigations of the origin of the modern humans from modern apes. Though the molecular phyllogeneticists and genomicists have scientifically revealed that the modern human species and modern apes (contemplated to be chimpanzee) shared a common ancestral lineage, from which the modern human species supposedly diverged out around 7-6 mya. So, sharing of such information encouraged the paleoanthropologists and paleontologists to look for fossilized ape-like creatures to identify and establish as the ancestral evolutionary model of apes from which the modern human species has diverged out. So far the anthropological investigation has successfully identified several pre-historic fossil apes through which the modern apes have been descended out. Hopefully, the scientists would be able to find out that enigmatic phylogenetic cousin, the unique fossilized ape that had been lost (disappeared in the trail of biological evolution) around 6-7 mya in the fourth dimension of time. A brief discussion on the origin and evolution of modern human species have further been narrated hereunder with the four distinct groups of Hominins (Pickrell, September 4, 2006; Pontzer, 2012; Menton, May 29, 2018):

A. Ardipithecus Group: The name *Ardipthecus* has been derived from the Afar Ethiopian word "Ardi" meaning ground and the Greek word Pithekos meaning Ape, which is representing the Ground Ape, or the first Ape that could able to walk or run upright posture. The evolutionists have considered *Ardipthecus* as the earliest ape-like animal of human ancestor, that looked like a chimpanzee, evolving around 5-6 mya in Ethiopia of the African continent. e.g. *Ardipithecus kadabba* and *A. ramidus* (Su, 2013).

Ardipithecus kadabba: In 1997, Yohannes Haile-Selassie, the great Ethiopian paleo-anthropologist had discovered this fossilized ape from Middle Awash, Ethiopia, who described it on the availability of its lower jaw. Later, 11 fossil specimens of *A. kadabba*, comprised of a different part of the body including leg, hand, arm, etc. have been described, anthropologically to belong to five individuals. The size of its brain resembled Chimpanzees and age-dating investigation determined it evolved around 5-6 mya (Su, 2013).

Ardipithecus ramidus: In 1994, Tim D. White, the eminent paleo-anthropologist from North America had reported *Ardipithecus ramidus* from the Middle Awash region of Ethiopia, Africa. From the initial investigation, the scientist had wrongly identified it as the species belonging to *Australopithecus* but a critical investigation of 100 pieces of fossil fragments, has rendered the scientists to recognize it as a new species *A. ramidus* under the newly formed genus *Ardipithecus*. The species name "*ramidus*" of *A. ramidus* has been derived from the Afar Ethiopian word "root," literally indicating this ape close to the root-stock of humanity. The age-dating investigation of evolutionary biologists indicated that it appeared around 4.4 mya and which indicated that it appeared around 1 million years before the emergence of "Lucy," the representative of Australopithecus (Su, 2013). The careful investigation of the foot bone and reconstruction of the pelvis of the fossil specimen indicated its arboreal grasping habit with occasional evidence of bipedalism from its big toe.

B. Australopithecus Group: The paleontological and paleoanthropological investigation in the human fossil site of Hadar, Afar region of the Great Rift valley of Ethiopia, Eastern Africa, had helped the scientists to discover a next group of hominin, Australopithecus, the prehistoric tree-climbing ape, that supposedly evolve around 2-4 mya. The word Australopithecus stands for Southern Apes, those have been contemplated by evolutionists that the members of Southern Apes have gradually adapted to walking in an upright posture (bipedalism) like modern human species, though they have anatomically, predominantly contended to be tree climbers and knuckle walkers (Tuttle, 1969; Johanson and Edey, 1981; Stern and Susman, 1983; Berge and Goularas, 2010). e.g. *A. afarensis*, *A. africanus*, *A. anamensis*, *A. sediba*.

Australopithecus afarensis: In 1974, a group of US paleoanthropologists, under leadership of Donald Johanson had discreetly studied the 47 fossil bones prehistoric hominin (of different parts of the body: ribs, vertebra, pelvis, part of skull, etc.) and reconstructed further to form 3.5 feet tall, earliest, female hominin fossil, "Lucy," recognized as "Mother of all humankind" and Lucy had been considered to belong to *Australopithecus Afarensis.* With its composite features like a typical ape-like skull and human-like pelvis, Lucy as well as *A. afarensis*, which has found to be evolved around 3.85 to 2.95 mya, have considered being the missing link between the last line of apes and the first line of humans (Johanson and Edey, 1981).

Australopithecus africanus: In 1924, Raymond Dart, the Australian born South African anthropologist had critically studied the carinal anatomy of the skull of the hominin fossil, discovered to claim the ape-like human ancestor from South Africa, popularly called "Taung child," the pre-historic fossilized hominin specimen of a juvenile (around 3-4 years old) and scientifically recognized as *Australopithecus africanus* (Ward and Hammond, 2016). The evolutionary age of it has been determined between 3.3 to 0.87 million years.

Australopithecus anamensis: In 1995, Meave G. Leakey, the British paleo-anthropologist and her fellow scientists had discovered the hominin fossils of teeth and bones from Kanapoi in Northern Kenya, that belonged to *Australopithecus anamensis*, evolving around 3.9 to 4.2 mya. The earlier paleoanthropological investigation from the same fossil site in 1965, has helped to discover the 4.5 million years old fossilized humerus (KNM-KP 271) and the critical observation of that fossil has rendered scientists like Henry M. McHenry to reveal that the fossilized hominin specimens from Kanapoi, found to be 4-4.5 million years old, was found to be anatomically indistinguishable from the modern human species, but too old to be considered a human fossil (McHenry, 1975).

Australopithecus sediba: In 2008, Lee Rogers Berger, North American born South African paleoanthropologist, along with his two sons, had discovered the fossil specimen of a hominin from Malapa cave, adjacent to Johannesburg, South Africa and commonly recognized it to be "Malapa Boy," scientifically called *Australopithecus sediba.* The evolutionary biologists and paleoanthropologists have determined the Malapa boy to have evolved around 1.78 to 1.95 mya, so the scientists have expressed their concerns to whether Malapa boy was found to be too young to qualify as an ancestor of the modern human species or not (Stern and Susman, 1983; Berge and Goularas, 2010)? The anatomical investigation of Malapa boy concerning its skull, skeleton, curved fingers, had rendered it a closer affinity to t hetree-dwelling, knuckle-walking apes to a greater extent and modern human species to a lesser extent (Ward and Hammond, 2016).

C. Paranthropus Group: This group of fossilized apes, belonging to Hominins, had formerly been assigned to the category of *Australopithecus* but careful observations from paleo-anthropologists has helped to segregate it and put in the category of *Paranthropus* group. The members of the *Paranthropus* apes were characterized by bigger sized teeth with enamel (identical to human teeth anatomy), bigger jaws with stout jaw muscles, where most of these features were absent or poorly found in the members of *Australopithecus*. When the members of *Australopithecus* had possessed composite features like predominating tree climbing arboreal behavior with occasional bipedalism, typical features similar to the modern human species, the members of *Paranthropus* group were contemplated to be drifted to the human-like behavior and adapted to bipedalism, came out from the typical feature like tree-climbing behavior (Arambourg and Coppens, 1968). Evolutionary biologists observed that it evolved around 1.2 to2.7 mya, almost at the same time that Australopithecus evolved e.g. *P.aethiopicus*, *P. boisei*, *P. robustus* (Herries et al, 2020).

Paranthropus aethiopicus: In 1985, North American paleoanthropologists duo Alan Walker and Richard Leakey, had discovered a fossil skull (characterized by sagittal crest, big jaw with large teeth and stout facial muscle), popularly known as "Black Skull," which has been scientifically known as *Paranthropus aethiopicus*, found in the fossil site of Turkana basin in Northern Kenya and Southern Ethiopia, in Eastern Africa. Evolutionists have contemplated that it supposedly evolved 2.7 to 2.3 mya (Arambourg and Coppens, 1968).

Paranthropus boisei: In 1959, Mary Leakey, British paleoanthropologist had discovered *Paranthropus boisei*, popularly known as Zinjanthropus from the fossil site of Olduvai George of Northern Tanzania. The features of Gorilla-like sagittal crest, big jaws, stout facial muscle, smaller carinal cavity like apes and human-like features like bipedalism and teeth with enamel, rendered this ape to be put in evolutionary transition between ape and human (Walker et al, 1986).

Paranthropus robustus: In 1938, Robert Broom, a Scottish origin South African paleoanthropologist had discovered the fossil fragments of apes (jaws, molar teeth, and some bones), popularly known as "Beside Man" from the fossil site of Kromdraii, South Africa. This prehistoric ape has been characterized by the dimorphic fossil specimen of 3 feet 9 inches - 3 feet 3 inches in height with Gorilla-like sagittal crest, sexual dimorphism, stout facial muscle like apes and human-like teeth with enamel (Broom, 1938).

D.Homo Group: Evolutionary investigation and palaeo-anthropological analysis along with the review of enormous hypotheses and relevant extrapolation had helped the scientists and the contemporary community of science that we the representatives, belonging to modern human species have got in touch with our prehistoric, phylogenetic ancestors, belonging to Hominins, scientifically recognized as Homo group. The cumulative investigation of paleoanthropologists along with paleontologists and phyllogeneticists helped to decipher that the earliest members, belonging to this group had emerged around 2.4 mya, that has found to be overlapping to the other extinct members belonging to Australopithecus and Paranthropus group in hominins, and the common features shared by humans was the presence of enamel in their teeth and bipedal mode of walking (van Holstein and Foley, 2017).

Homo erectus: In 1891, the earliest fossilized human skull, which is popularly recognized as "Java Man," scientifically known as *Homo Erectus* had been discovered by the Dutch Anatomist Eugene Dubois and the earliest fossil specimen of H. erectus, popularly known as "Peking Man" had been discovered from the fossil site of Zhokoudian, nearby Beijing, China, around 200 fossil specimens (belonging to 40 individuals) of which have been investigated by the Canadian paleoanthropologist Davidson Black, and the evolutionary age of it has been determined to be between 1.9mya and 144,000 years ago. In 1984, Richard Leakey, the North American paleoanthropologist had discovered another fossil specimen of *H.erectus* from Lake Turkana in Northern Kenya, popularly known as "Turkana Boy." The key features like carinal cavity of this complete fossil, which is around 800cm3, typical mode of bipedal walk (evidenced by anatomical formation of Pelvis) and the presence of some tools like hand axes, fire hearths, etc. rendered the scientists to contend it as a phylogenetically close ancestral lineage of the modern human species (van Arsdale, 2013).

Homo floresiensis: In 2003, a hominin fossil (around 3 feet 6 inch long) has been discovered by the paleoanthropologists from Island of Flores, Indonesia, popularly known as "Hobbits." Due to its bone deformation (either due to the disease of microcephaly or Down's Syndrome), the scientists have failed to make any unanimous declaration for Hobbits to be scientifically recognized as *Homo floresiensis.* The age-dating investigation of this skeleton and stone tools, adjacent to it revealed it existed around 50,000 to 190,000 years ago and also helped it to be recognized as the most recent ancestral lineage of modern human species (Ritter, 2016; Sutkina et al, 2016).

Homo habilis: In 1964, a team of three British-South African paleo-anthropologists, comprised of Louis Leakey, Phillip V. Tobias, and John R. Napier had discovered the fossil remnants of skull, hand and leg bones of four juvenile hominins, from the fossil site of Olduvai Gorge, in Tanzania. The scientists found that the hand and foot bones are identical to humans but skull fossil closely resembled apes. As scientists had noticed some stone tools nearby the fossil sites, they recognized *Homo habilis* as "handyman." As the fossilized skull of *Homo habilis* had not been found close to its hand and feet bones, some primatologists have considered it as the mixture of fossil specimens of *Homo erectus* and *Australopithecus* (Wood, 2014).

Homo heidelbergensis: The discovery of archaic humans under the genus Homo had been discovered by a fossilized mandible of *Homo heidelbergensis* from the fossil site of Heidelberg, Germany, for the first time by the German paleo-anthropologist Otto Schoetensack in 1907 (Buck and Stringer, 2014). The critical observation by the scientists have ensured that its skull had possessed common features of *Homo erectus* and the modern human species and the carinal cavity of it found to be as same or bigger than the modern human species, rendering it a speculation among scientists whether *H. heidelbergensis* was more intelligent than the modern human species or *Homo sapiens* or not (Lieberman et al, 2002) ? Since 1997, a series of primatological investigation ascertained that *H. heidelbergensis* evolved and diversified in different parts of the world ranging from England, France, Greece, Hungary, Italy, Portugal, Spain, Ethiopia, Namibia, South Africa and they have roamed around 700,000 to 200,000 years ago (Mounier et al, 2009).

Homo neanderthalensis: In 1856, a group of German paleoanthropologists has discovered the fossil specimen of Neanderthals in the Feldhofer Cave of Neander valley in Germany, though the erstwhile fossil explorations in Engis, Belgium and 1848 at Forbes Quarry, Gibraltar had successfully discovered it earlier. So far scientists have found the fossils of Homo neanderthalensis from the Eurasian region but they did not get it from Africa, yet age-dating techniques ascertained it existed around 400,000 to 40,000 years ago. Paleoanthropologically (presence of evidence like hand-tools, artifacts, musical instruments, jewelry, costumes, knowledge of medicine, fire-hearths, seafaring, etc.), Neanderthals were found to be culturally, technically, behaviorally comparable to the modern human species. This immediate phylogenetic ancestor of modern human species was either exterminated by migrating population of the modern human species, or by climate changes or due to contraction of the pathogenic disease, that rendered it extinct (Banks et al, 2008; Underdown, 2008; Wolf et al, 2018).

Homo rudolfensis: In 1986, Richard Leakey, the Kenyan paleoanthropologist has discovered the fossilized skull (KNM-ER 1470) from the fossil site of Koobi Fora, nearby Lake Turkana, Northern Kenya. Though the skull with 800 cm3 carinal cavity had maxilla, and swung forward in direction seems to appear like apes, the critical investigations by scientists that noted the lack of any trace of brow and sagittal crest have rendered it to have human-like appearance and age-dating has indicated its evolution to supposedly taken place around 1.9 mya (Lewin, 1984).

Homo sapiens: Based on paleontological and molecular evidence, about molecular genetics and genomics, the paleoanthropologists and evolutionary biologists have contemplated that the anatomically modern human species, the extant human on Earth had evolved around 200,000 years ago in Africa (Cann et al, 1987). In 1857, the eminent Swedish taxonomist had introduced the scientific name of this extant, modern human species as "*Homo sapiens*," scientifically, which stands for "wise man," structurally lesser robust and appeared with prominent brow ridges on its forehead in respect to other members of Homo, the other extinct members, belonging to Hominins. Several paleontological evidence of modern human species had been found in different times of exploration, some of those are as follows (Smith and Ahern, 2013):

Cro-magnon man from Europe
Denisovan from Siberia
Grimaldi man from Italy
Piedmont from the Czech Republic etc.

Though scientists have come to a unanimous conclusion that there is one human genus "*Homo*," one species "*sapiens*" and subspecies "*sapiens*," found to be in extant state on Earth but there is no substantial evidence that we could refer with 100% certainty that any particular extinct species of "Homo," would be recognized further as ancestor of *Homo* sapiens and there is no first-hand evidence that any particular species of *Homo* might have been evolved out of another species in course of journey through the trail of evolutionary history of humanity. So we are still in the labyrinth of the conjecture of the history of the origin of the modern human species, where, logically, Darwinian doctrine of evolution of Human evolution, which has been narrated in "*The Descent of Man*" in 1971 and has been left for its readers to exercise his or her prejudices to get integrated to a larger perception of origin and evolution of humanity as the evolutionary analysis is still going through the labyrinth of conjectures, whereas the Biblical interpretation, as well as the interpretation of special creationism to contemplate Adam and Eve as the original couple of humanity, seems to be good theological proposition (Genesis 1-5), who supposedly initiated the journey of humanity from the "Garden of Eden." According to Jean-Jacques Hublin, the present director at the Max Planck Institute for Evolutionary Anthropology in Leipzig, Germany and the leading paleo-anthropologists of the fossil excavation site of Jebel Irhoud, Morocco, at the Atlantic coast engaged in the discovery of earliest specimen of *Homo sapiens* ascertained that modern human species had evolved 315,000 years ago (whereas most of the paleoanthropologists have found its emergence somewhere from east Africa around 200,000 years ago) stated "*Until now, the common wisdom was that our species emerged probably rather quickly somewhere in a 'Garden of Eden' that was located most likely in sub-Saharan Africa,*" he further said, "*I would say the Garden of Eden in Africa is probably Africa — and it's a big, big garden.*" (Callaway, 2017).

References

Agin, D. (2006). Junk Science: How politicians, corporations, and other hucksters betray us. Thomas Dunn Books. P. 336.

Arambourg, C., and Coppens, Y., (1968). Sur la decouverte dans le Pleistocene inferieur de la valle de l'Omo (Ethiopie) d'une mandibule d'Australopithecien. *Comptes Rendus des seances de l'Academie des Sciences* 265: 589-590.

Banks, W.E., d'Errico, F., Peterson, A. T., Kageyama, M., Sima, A., and Sánchez- Goñi, M. F. (2008). Neanderthal extinction by competitive exclusion. *PLoS One*. 2008;3 (12): e3972. doi: 10.1371/journal.pone.00039 72. Epub 2008 Dec 24.

Berge, C., and Goularas, D. (2010). A new reconstruction of Sts 14 pelv is (*Australopithecus africanus*) from computed tomography and three-dimensional modeling techniques. *J Hum Evol.* 58(3): 262-72. doi: 10.1016/j.jhevol.2009.11.006.

Broom, R. (1938). The Pleistocene anthropoid apes of South Africa. *Nature* 142: 377-379. doi:10.1038/142377a0

Buck, L.T., and Stringer, C. B. (2014). *Homo heidelbergensis*. *Curr Biol.* 24(6): R214-5. doi: 10.1016/j.cub.2013.12.048.

Bushfield, A. (March 4, 2018). Chimps believe in God, researcher claims.

https://premierchristian.news/en/news/article/chimps-believe-in-god-researcher-claims

Callaway, E. (2017). Oldest *Homo sapiens* fossil claim rewrites our species' history. *Nature* (2017). https://doi.org/10.1038/nature.2017.22114

Cann, R.L., Stoneking, M. and Wilson, A.C. (1987). Mitochondrial DNA and human evolution. *Nature*. 325: 31-36.

Denton, M. (2002). *Evolution: A theory in crisis*. (3rd Edition.)Adler & Adler. P. 368.

Hayden, T. (February, 2009). What Darwin didn't know?

https://www. smithsonianmag.com/science-nature/what-darwin-didnt-know-45637001/

Hendel, R.S. (January 08, 2021). The search for Noah's flood. Scientists are looking in the wrong place.

https://www.biblicalarchaeology.org/ daily/biblical-topics/hebrew-bible/the-search-for-noahs-flood/

Herries, A.I. R., Martin, J.M., Leece, A.B., Adams, J.W., Boschian, G., Joannes-Boyau, R., Edwards, T.R., Mallett, T., Massey, J., Murszewski, A., Neubauer, S., Pickering, R., Strait, D.S., Armstrong, B.J., Baker, S., Caruana, M.V., Denham, T., Hellstrom, J., Moggi-Cecchi, J., Mokobane, S., Penzo-Kajewski, P., Rovinsky, D.S., Schwartz, G.T., Stammers, R.C.,

Wilson, C., Woodhead, J., Menter, C. (2020). Contemporaneity of *Australopithecus*, *Paranthropus*, and early *Homo erectus* in South Africa. *Science*. 368(6486): eaaw7293. doi: 10.1126/science.aaw7293.

Hunt, K.D. (2016). Why are there apes? Evidence for the co-evolution of ape and monkey ecomorphology. *Journal of Anatomy* 228(4): 630-685.

Johanson, D. C. and Edey, M. A. (1981). *Lucy: The beginnings of humankind.* New York, NY: Simon and Schuster. P. 416.

Kundu, S. R. (2018). *Origins and Evolution of Plants on the Earth and the Descendants of ANITA*. Nova Science Publishers Inc. p.387.

Lewin, R. (1984). *Human Evolution: An illustrated introduction.* Blackwell Publishing. p.228.

Lieberman, D.E., McBratney, B.M., and Krovitz, G. (2002). The evolution and development of cranial form in *Homo sapiens*. *PNAS USA* 99: 1134-1139.

McHenry, H. M. (1975). A new pelvic fragment from swartkrans and the relationship between the robust and gracile australopithecines. *American Journal of Physical Anthropology*. 43(2): 245-261.

Menton, D. (February 25, 2010). Did humans really evolve from Apelike creatures?

https://answersingenesis.org/human-evolution/ape-man/did-humans-really-evolve-from-apelike-creatures/

Menton, D. (May 29, 2018). Differences between humans and apes. Featured in "Searching for Adam."

https://answersingenesis.org/adam-and-eve/did-humans-really-evolve-from-ape-like-creatures/

Millman, J., Taylor, B., and Effron, L. (December 5, 2012). Evidence Noah's Biblical flood happened, says Robert Ballard.

https://abcnews.go. com/ Technology/evidence-suggests-biblical-great-flood-noahs-time- happened/story?id=17884533

Mounier, A., Marchal, F., and Condemi, S.(2009). Is *Homo heidelbergensis* a distinct species? New insight on the Mauer mandible. *J Hum Evol.* 56(3):219-46. doi: 10.1016/j.jhevol.2008.12.006. Epub 2009 Feb 27.

Peppe, D.J. and Deino, A.L. (2013). Dating Rocks and Fossils Using Geologic Methods. *Nature Education Knowledge* 4: 1

Pickrell, J. (September 4, 2006). Timeline: Human Evolution.

https://www. newscientist. com/article/dn9989-timeline-human-evolution/

Pontzer, H. (2012). Overview of Hominin Evolution. *Nature Education Knowledge* 3(10):8.

Radford, T. (April 25, 2011). The untold story of evolution.

https://www.theguardian.com/science/2011/apr/25/evolution-human-history-apes

Riley, B. (December 30, 2008). Our Relatives, the Dinosaurs.

https://www. smithsonianmag.com/science-nature/our-relatives-the-dinosaurs- 36503129/

Ritter, M. (30 March 2016). "Study: Indonesia "hobbit" fossils older than first thought". https://apnews.com/article/59d725f16f2945b08e5ee9f0b8c2c286

Rocatti, G., and Perez, S. I. (2019). The Evolutionary Radiation of Hominids: a Phylogenetic Comparative Study. *Sci Rep.* 9(1): 15267. doi: 10.1038/s41598-019-51685-w.

Smith, F.H. and Ahern,J.C.M. (2013). *The Origins of Modern Humans: Biology Reconsidered.* John Wiley & Sons Inc. p. 454.

Smithsonian National Museum of Natural History (March 15, 2010). What does it mean to be human? https://humanorigins.si.edu/evidence/human-fossils

Stern, J.T., and Susman, R.L. (1983). The locomotor anatomy of *Australopithecus afarensis*. *Am J Phys Anthropol* 60: 279–317.

Su, D. F. (2013). The Earliest Hominins: *Sahelanthropus*, *Orrorin*, and *Ardipithecus. Nature Education Knowledge* 4(4):11.

Sutikna, T., Tocheri, M.W., Morwood, M.J., Saptomo, E.W., Jatmiko, Awe RD, Wasisto, S., Westaway, K.E., Aubert, M., Li, B., Zhao, J.X., Storey, M., Alloway, B.V., Morley, M.W., Meijer, H.J., van den Bergh, G.D., Grün, R., Dosseto, A., Brumm, A., Jungers, W.L., and Roberts, R.G. (2016). Revised stratigraphy and chronology for *Homo floresiensis* at Liang Bua in Indonesia. *Nature.* 532(7599): 366-9. doi: 10.1038/nature17179. Epub 2016 Mar 30.

Trinkaus,E. (1985). Cannibalism and burial at Krapina. *Journal of Human Evolution* 14(2): 203-216.

Tuttle, R. H. (1969). Knuckle-walking and the evolution of Hominoid Hands. *Am J Phys Anthropol* 26: 171– 206.

Underdown, S. (2008). A potential role for transmissible spongiform encephalopathies in Neanderthal extinction. *Med Hypotheses*. 71(1):4-7. doi: 10.1016/j.mehy.2007.12.014. Epub 2008

van Arsdale, A. P. (2013). *Homo erectus* - A Bigger, Smarter, Faster Hominin Lineage. *Nature Education Knowledge* 4(1):2.

Ward, C. V. and Hammond, A. S. (2016). *Australopithecus* and Kin. *Nature Education Knowledge* 7(3):1.

Wolf, D., Kolb, T., Alcaraz-Castaño, M., Heinrich, S., Baumgart, P.,Calvo, R., Sanchez, J., Ryborg, K., Scafer, I., Bliedtner, M., Zech, R., Zoller, L. and Faust, D. (2018). Climate deteriorations and Neanderthal demise in interior Iberia. *Sci Rep* 8: 7048. https://doi.org/10.1038/s41598-018-25343-6

van Holstein, L., and Foley, R.A. (2017). Hominin Evolution. In: Shackelford T., Weekes-Shackelford V. (eds) *Encyclopedia of Evolutionary Psychological Science*. Springer, Cham.

https://doi.org/10.1007/978-3-319-16999-6_3416-1.

Wang, X., Grus, W.E.,and Zhang, J. (2006). Gene Losses during Human Origins. *PLoS Biol* 4(3): e52. https://doi.org/10.1371/journal.pbio.0040052

Walker, A.C., Leakey, R.E., Harris, J.M., and Brown, F.H., (1986). 2.5-Myr *Australopithecus boisei* from west of Lake Turkana, Kenya. *Nature* 322: 517-522.

Wilson, E. O., and Peter, F. M. (1988). *Biodiversity*. Washington, D. C., National Academy Press. P.521.

Wood B. (2014). Human evolution: Fifty years after *Homo habilis*. *Nature*. 508(7494):31-3. doi: 10.1038/508031a.

Young, N. M., Capellini, T. D., Roach, N. T., and Alemseged, Z. (2015). Fossil hominin shoulders support an African ape-like last common ancestor of humans and chimpanzees. *PNAS U S A*. 112 (38): 11829-34. doi:10. 1073/ pnas.1511220112. Epub 2015 Sep 8.

Chapter 4

Origin of Modern Human Species in the Evolutionary Labyrinth of Adam, Eve & Apes: II (Peeping through Molecular Evidence)

In contrast to Great Apes, who have 48 chromosomes, modern humans and likely Neanderthals and Denisovans have and had, respectively, 46 chromosomes. The reduction in chromosome number was caused by the head-to-head fusion of two ancestral chromosomes to form human chromosome 2 (HSA_2) and may have contributed to the reproductive barrier with Great Apes. Next-generation sequencing and molecular clock analyses estimated that this fusion arose prior to our last common ancestor with Neanderthal and Denisovan hominins ~ 0.74 - 4.5 million years ago.

---- *Pawel Stankiewicz*

Whether the evolutionists would claim the emergence of the modern human species as a derived form, which has been descended from an ancestral root-stock of ape, and the creationists have taken a firm stand that we have been created as the ultimate wish of God, that defends our miraculous creation on Earth, the perceptional conflict of our emergence on Earth could have remained as a unique matter of scientific or theological mystery for future generations but the reality is that we the modern human species have emerged as an ideal form of a social animal rather than other biological species including apes, even if we consider them as our evolutionary predecessor. In any society, theological influence played the cardinal role to set and roll-out several socio-religious laws, socio-behavioral rules, protocol, socio-cultural norms, and practices that the members of the society follow during their inter and intra-societal communications. It is difficult to comprehend that (even in any society, under the socio-political influence of communism) there is any existence of any human society, where the social norms and rules have been controlled and influenced by atheists and theological practice, and perceptions have no place in the life of anyone of that society. So in general, any society of the modern human species, most of the individuals either born as God-loving or God-fearing people, even the radical supporters of the biological evolution have mostly emerged out of this society and gradually drifted away from the perceptional arena of special creationism to get into the arena of science, specifically the broader podium of biological evolution. Let's start with the Biblical narrative of human ancestry, where Genesis 1-2 clearly defines that God created man in His image and by His power of special creation, He formed Adam out of dust and formed Eve from Adam's side and that is clear enough to depict man as a distinct kind of entity, should not contend any common identical features (either partial or absolute) with other animals. The Biblical narrative of the ancestry of Humanity had started with the depiction of Adam and Eve as the maiden progenitors of the modern human race, especially Eve who had been introduced as the mother of all living (Genesis 3:20) and introduced Noah, his wife, their three sons and their wives (that's eight people on the Ark) that contended to the ancestral population of the modern human race (the whole earth was populated, Genesis 9:19). In the Biblical narrative, it has further been narrated immediately after the commencement of Noah's Flood, the emergence and formation of the ancestral lineage of modern human "race" or ethnic groups had been taken place as a confusion of languages in the Tower of Babel (Gen 11:8-9) (Jeanson, 2017, 2019). So, the Biblical narrative of the emergence of humans is pretty straightforward, but that does not necessarily mean that such interpretation of special creation of humanity is acceptable to everyone ranging from orthodox creationists to radical evolutionists or even a group recognized as Evolutionary creationists or theistic evolutionists. So let's go for a brief review of another hypothesis to get a feasible depiction of the origin of humans on Earth. According to the YEC (young-earth creation) hypothesis, promulgated by the theistic evolutionists, the emergence or creation of Adam and Eve had taken place around 6000 years ago, whereas the historical population bottleneck in the background of "Flood of Noah" occurred around 4500 years ago and the occurrence of confusion of languages in the Tower of Babel had supposedly taken place after the flood incident (Hardy and Carter, 2014; Carter and Hardy, 2015). But, if we try to interpret the evolution of humanity from the perspective of evolutionists, the initial Darwinian perception of human evolution was initiated with an absolute denial of the creationists about human evolution as evolutionists have firmly stuck with an observation, which claimed that apes are our close relatives (The Chimpanzee Sequencing and Analysis Consortium, 2005). Contemporary phylogenetic research had revealed that the Chimpanzee was the nearest phylogenetically related to humans and the split of these two distinct lineages took place around 3-13 mya (Langergraber et al, 2012).

It is convincible that as a distinct model, evolution and YEC have tended to define the origin of humans in a unique way, where the prehistoric humans and apes might not be fitting in identical slots according to evolutionary narratives or the alternative narratives of YEC, but it does not necessarily mean that either model is flawless or erroneous. From the perspective of basic science (anatomical and genetic perspective), apes are considered to be identical to humans. So, there should be an evolutionary trend or phylogenetic consensus that exists among Gorilla, Chimpanzee, Orangutan and modern human species that supposed to be expressed by morphological features of similarities or dissimilarities but that kind of trend has hardly been witnessed by the evolutionists. Rather, the molecular phyllogeneticists have shared their unique research observation and ascertained that they have contemplated Chimpanzee as the nearest phylogenetically of the modern human species (Grehan and Schwartz, 2009).

So, in her blog, entitled "How do Adam and Eve fit with evolution?" when Stacy A. Trasancos stated, "Biological evolution will never fully account for humanity because we are persons, made in the image and likeness of God." (Trasancos, March 28, 2017). To elucidate the logical shortcomings of biological evolution in defining unanimous acceptance of the hypothetical perception of the emergence of the modern human species, Dr. Trasancos stated "Tell a Catholic kid about evolution—that there was a Big Bang and that in this expanding cosmos our sun is a star in a cluster of 200 billion stars in the arm of a spiral in a galaxy among thousands, and that eventually on our planet there appeared early life forms, single-celled bacteria, trilobites, mollusks, fish, amphibians, reptiles, birds, mammals, primates, and that from a common ape-like ancestor the Homo genus emerged, including Homo rudolfensis, Homo habilis, Homo erectus, Homo heidelbergensis, Homo floresiensis, Homo neanderthalensis, and last, Homo sapiens, or "wise man," the hominin species that is modern human." and she mentioned "The very next question that kid will ask is, "So where do Adam and Eve fit in?" (Ask me how I know!) It is a logical question." (Trasancos, March 28, 2017). The logical interpretation of the emergence of the modern human species, mentioned above, seems to be one of the best logical and scientific interpretation of the evolution of humans but that was found to be challenging and difficult to be established to some-one when and if someone was found to be strongly integrated to a society having a strong sense of theological orientation and don't just consider him or herself as a representative of a biological species, rather than be recognized as a superior biological entity with limitless unlocked potential in this regime of Anthropocene. To get a feasible solution of any hypothetical concern, the best way is to sort the fundamental question out that would address the problem in a straight way, without intimidating the atheist evolutionists or creationists, belonging to Abrahamic religious faith group. As most of the religious institutions of any society could deal with the opposing religious group (like recognizing God as Ronen instead of Ramos) if it's necessary but they are hardly ready to compromise with atheist evolutionists (no existence of God) under no circumstances. According to Dr. Trasancos, the best way is to maintain a state of equilibrium is simply admitting "We do not know the exact biological details of Adam and Eve, and we never will" and she further elucidate that once the framing of the main question has been set, out of any biased perception, the analogical interpretation of any complicated hypothetical issue seems to be easier to comprehend, through the narratives of creationism rather than a scientific interpretation of evolution. Dr. Trasancos has tendered a narrative of two ancestral grains of sand, considered to be the maiden one that formed that beach, and scientifically it is an almost impossible task to narrow it down, that gave rise the entire beach of sand through successive cycles of sand formation but same issues could have been feasibly answered in the perspective of the theological perspective of creationism. Likewise, tracing out a primordial couple and designate them as the root-stock of humanity is not only an impossible issue to the evolutionists as evolution has normally deal with a population (even if it's a scanty one) to track down the evolutionary history of that species, no feasible evolutionary model is there that could narrate the evolutionary history of any species (like modern human species), historically (as well as theologically) contemplated to be descended out of a primordial couple. Even the availability of the paleontological evidence, that intriguing the scientists to go for radiometric dating of that sample, in the backdrop reference of geological time-scales of thousands or millions years for remote past so determining the identity or finding out a primordial couple of Adam and Eve as the origin of Humanity seems to be an almost impossible agenda of science to trace out two maiden grains of sands of ancestral origin out of the entire beach of the sand.

But the closing of one way does not necessarily mean we are on the dead-end of the enigmatic issues like origin of humanity as the closing of one easier way often open up of an alternative (maybe difficult one) way to get closer to a feasible answer of the difficult question of this complicates issues. However, it has been mentioned earlier that a world-wide investigation on the mitochondrial DNA utilized molecular clock ascertained that most or all mitochondrial DNAs supposedly emerged out of one primordial super-woman, Mitochondrial Eve, who lived in Africa around 200,000 years ago (Cann et al, 1987). Likewise, it has been mentioned earlier that Y-chromosomal Adam supposedly lived around an identical period (Eck, August 3, 2013; Francalacci et al, 2013). However, the scientists would not confirm that both of them lived together in the same place and we are offsprings of that monogenetic pair as well. Rather scientists logically argued in favor of a bottleneck of a small pre-historic human population in the time where the population size would have been considered to be around 10,000 and Trasancos (March 28, 2017) have further contended that such findings in the field of human phylogenetic investigations are provisional, and the possibility of better calibration would inevitably render better results with more precision. The most important thing is that sharing or promoting such observation does not hurt the evolutionists, and doesn't patronize the spirit of creationists, so it creates an opportunity to reconcile the theological perception to present it uniquely in the scientific perspective of an evolutionist's vision. However, several scientists, especially the molecular biologists having expertise on a computer algorithm and modeling on genealogical computation, found out that vertical backward or cascading of computation of the most recent representative of modern human species would help get in touch with the virtual progenitor of us, that existed on Earth around few thousand years ago (Rohde et al, 2004). And further refinement of data with the precise effort of computation cascading would help us to come across our maiden pre-historic human population, that lived on Earth between 50,000-200,000 years ago (Noonan, 2010).

Since the nineties, the contemporary studies of molecular foot-print of life, especially in deciphering the unsolved phylogenetic mystery like origin and evolution of a certain group of organisms, like modern human species. Let's go back to square one, to have a brief review of the evolutionary perception of the evolutionary biologists like Grehan and Schwartz (2009), who contended to project humans as a genetically identical and phylogenetically of the chimpanzee as well. The acceptance of any hypothesis or scientific model depends on the match of predictions and the facts/results and discrepancies between this two-end rendering rejection of that experimental model or hypothesis. So the human-ape hypothesis of the origin of the modern human species as well as the genetic comparison between human and chimpanzee has been logically reviewed and elucidated simply by Dr. Nathaniel T. Jeanson alongwith his research colleagues (Jeanson, September 14, 2019; Purdom et al, March 1, 2019), to find out whether predictions fit the facts of the evolutionary model of the origin of human from apes.

Furthermore, Jeanson (September 2019; April 25, 2021) had referred to the text of a genetic book under the biological system, specifically in our cell, contained with biochemical substance, called DNA and he further referred it as functional or operation manual of our cell. As the biochemical instructions have found to be encoded in a 4-letter chemical alphabet and the combinations of these letters rendered the formation of words and sentences that carry biological messages from one generation to the next generation. However, the scientists have figured out that the DNA in our cell contained billions of letters long, has been contemplated as a large biological book (Mortenson, 2014; Jeanson, September 14, 2019; April 25, 2021). The reproductive biologists have revealed that before conception, the copying of DNA, taken place in egg or sperm, found to be erroneous (due to copying mistake) and which is technically called a mutation. The evolutionary biologists have observed that the rate of mutation (it's around 60 mutations happen in each human generation) happened during the copying of DNA in humans and chimpanzee found to be identical (Conrad et al, 2011; Venn et al, 2014). As scientists have revealed earlier that modern human species and chimpanzee have split out from the common root-stock around 3 to 13 mya and the average life span (i.e. duration of each generation) from one generation to next took 20 years, it has been contemplated from a feasible back-tracking that around 150,000-650,000 [3 to 13 million years/20 years] generations ago these two species, supposedly diverged out of a common ancestor (Mortenson, 2014). A further extrapolation of it has found to be quite interesting as at the rate of 60 DNA mutations per human generation (during copying) in the generation of 150,000 to 650,000, the number of DNA changes would be in the range of (150,000 to 650,000) x 60 or 9 million to 39 million DNA. As the same process has been repeated for Chimpanzee too we need to double it up to narrate a tentative assumption that DNA of humans and DNA of the chimpanzee supposedly differ around 18 to 79 (~80) million DNA letters (The Chimpanzee Sequencing and Analysis Consortium, 2005; Mortenson, 2014; Wong, September 1, 2014). Jeanson (April 25, 2021) has tried to figure out the empirical difference of DNA, in a unique way he has tendered an analogy of determining the number of pages by counting the number of letters and words (average number of letters per word and or number of words per page of the book concerned) and he defined further that the total letter count in chimpanzees and humans is calculated (6 billion DNA letters of diploid [2n] species of human and chimpanzee), to be around 3 billion DNA letters for haploid [n] species. The hypothetical prediction of the evolutionists about the contemporary genetic difference (difference of DNA) between human and chimpanzee assumed to be [78,000,000 DNA difference/ 3,000,000,000 letters of DNA in human = 0.026 or 2.6 % difference of around ~3%] 1-3% (DeWitt, January 14, 2014; Jeanson, April 25, 2021). Though the predictive value of the genetic difference (actually difference of DNA) between chimpanzee and human species is assumed to be in 1 to 3 % range the investigative studies of Jeffery Tomkens (December 28, 2011), entitled "*Genome-Wide DNA Alignment Similarity (Identity) for 40,000 Chimpanzee DNA Sequences Queried against the Human Genome is 86–89%*," published in the *Answers Research Journal*, has revealed further that the real difference has gone up to 12% level (i.e. 10 % above the level of prediction). Dr. Matthew W. Mitchell and his research colleagues's (2015) independent level of investigation has revealed that the human genome contains 3,259,520,000 DNA letters whereas the genome of chimpanzee contains 3,309,000,000 DNA letters (Mortenson, 2014). At the rate of 12% DNA difference from apes, the numerical difference of DNA letters between chimpanzee and humans would be (3,259,520,000 DNA letters of the human) around 0.12 x 3,259,520,000 DNA letters or 391,142,400 (~ 400 million) DNA letters difference (Baltimore, 2001; Tomkens, December 28, 2011; DeWitt, January 14, 2014). So, it has been observed by several scientists that the evolutionary prediction did not match to the fact in a strict sense. Naturally, it would not justify the observation like the difference between human and chimpanzee supposedly found to be around 1% (as predicted by several evolutionists that chimpanzee and human were genetically 99% identical to each other) as the precise molecular investigation ascertained that humans and chimpanzees were found to be genetically 88% identical and that rendered to invalidate the hypothetical origin of the human species out of apes, based on the mismatch of prediction to the fact; and nullify the evolutionary hypothesis of the origin of human from apes, rather than accepting it (Jeanson, September 14, 2019). In terms of quantitative index of genetic difference, when humans and chimpanzees were contended to be 88% identical these two species were differed by 400 million DNA letters and such result of the relative pattern of genetic differences between apes and humans, and humans and other lower group of vertebrates and a lower group of vertebrate to other terrestrial invertebrates and so on would promote the evolutionary hypothesis of ape ancestry in human evolution, as well as upholding the natural selection-driven "tree-like pattern of genealogical relationship among the living species" where the comparative account of genetic difference among apes, humans and other primates would have been diagrammatically presented by tree-like graphical presentation (Dendrograms) (Jeanson, September 14, 2019). So, we got a lesson that the predictive level of genetic difference between chimpanzee and human which was around 1 %, went up to 12% level in reality so the level of prediction did not fit the fact, and rendered the hypothesis invalidate. In justifying the validity of any evolutionary hypothesis, less priority needs to be given on quantitative prediction over qualitative prediction to promote a model of nested hierarchy and Nathaniel Jeanson has contended further "*...if humans have ancestry before the first Homo sapiens, then evolutionists expect humans to be relatively close genetically to the great apes, then slightly less close genetically to the rest of the primates, then even less similar genetically to other mammals, and quite different genetically from invertebrates and plants*." But it might not be so straight forward the interpretation in the domain of evolutionary biology to validate or discard any hypothesis. Let's get into a simple analogy of transport system suppose if we compare a 4 wheeler car, manufactured by Tata and a 4-wheeler car manufactured by Mitsubishi, there would be a quantitative index of difference between the two cars (according to some parameters like fuel efficiency, mileage, speed, etc.) would be X, relatively small figure, if we would engage in comparing those to a pick-up truck, manufactured by Ford, the quantitative index of difference would be certainly noticeable (relatively bigger index than the index of difference between Tata and Mitsubishi car) and if those cars would be compared to a big commercial truck or trailer, the index of difference would certainly be remarkably high in comparison to the first instance and the relative difference among all automobiles validates and form a template reference that we could compare a different group of automobiles to refer how similar or different they are from each other. It is a simple analogy that evolutionary hierarchy works in the biological world where the phylogenetic relationship (either distant or close) exists among phylogenetically related animals or plants maintained ever. Now back to the first hypothetical situation where the index of difference between the 4 wheelers of Tata and Mitsubishi reported to be X. If those automobile companies engaged in manufacturing/upgrading new models of 4-wheelers and the new difference would be Y (different quantitative index than previous index X), did it invalidate the utility of reference template that helped to understand the functional and structural difference of different models of automobiles based on presence or absence of a set of similar or dissimilar characteristics? Likewise, ensuring validity or discarding the same to put an invalid tag on it based on a straight-forward observation of the discrepancies (between prediction and fact or observation) of a quantitative index of genetic difference is not found to be the right approach as judging any hypothetical model, specifically in the field of biological evolution, where an investigation has been involved with a series of Spatio-temporal extrapolation (due to the long period, ranging from thousand to million years and dynamic changes of space due to ever-changing nature of environment) and the consideration of some variables would play a cardinal role to yield a varied outcome of the quantitative index under the distinct experimental protocol, the empirical result should primarily be taken into consideration, rather than sticking to the changing figure of a quantitative index like the genetic difference between modern human species and chimpanzee found to go up in 12% level, as observed/calculated earlier, which was around 1 to 3% level and the evolutionary time of divergence between these two has been recalibrated from 3 to 6mya to 13 mya (Langergraber et al, 2012; Venn et al, 2014). The changes of the quantitative index in the experimental result, about the close evolutionary relationship between humans and chimpanzee or the evolutionary hypothesis of the origin of humans out of apes, would indicate that the concerned evolutionary model requires modification but it would not invalidate it logically. Apparently, the expectation of evolutionary view must not be identical to the expectation of YEC view but the theist evolutionists had recognized the "nested hierarchical patterns" as a common podium and they have contended further that the "*relative hierarchy of genetic differences among humans, great apes, mammals, and invertebrates fits the YEC model at least an evolutionary one. So, to claim nested hierarchical patterns in the biological world as exclusive evidence of evolution would be analogous to claiming that the existence of people proves YEC*" (Jeanson, September 14, 2019).

In support of the hypothesis of the emergence of the human lineage from ancestral root-stock of apes, evolutionists assumed that the extant human chromosome pair (the chromosome exist in pair in any living organism, that they inherit each from their biological parents) number 2 or HSA_2 had been formed as an accidental fusion of two ancestral chromosomes of apes (IIp and IIq), that the modern human species had inherited from the ancestral lineage of apes (Stankiewicz, 2016). So, the main concern was that modern human species have 23 pairs of chromosomes but the prehistoric apes (like chimpanzees, Bonobos, gorilla, etc.) have 24 pairs of chromosomes and the evolutionists have contemplated that in the apes chromosome a pair of ancestral chromosomes have supposedly fuse with another pair of the ancestral chromosome to emerged with a new form of human species with 46 pair of chromosomes, followed by a series of morph-anatomical and behavioral changes to be emerged as modern human species (Tijo and Levan, 1956; Jeanson, September 14, 2019).

Till, the 1970s, the progress of molecular techniques in DNA analysis was not that level that helped the evolutionary biologists to define chromosomal rearrangements and distinguish human species from apes, genetically. A unique investigation on chromosome 2 of modern human species by Lido et al (1991) revealed chromosomal rearrangement in the middle of chromosome 2, having telomeric and sub-telomeric DNA sequence, that likely presented that how are chromosome 2 got emerged as a result of the fusion of two chromosomes of the apes, that made the evolutionists to construe ape our phylogenetic successor. It needs to be mentioned herewith that the previous contemplation about the fusion of two ancestral chromosomes and the emergence of chromosome 2 in the Human genome followed by inactivation of one of the two centromeres has been reported by Yunis and Prakash (1982). Later, it had been observed by the molecular biologists that the DNA sequences, once found to be nested in the terminal end of the ancestral chromosome, have found to be located in the interstitial position (almost in the middle) of human chromosome number 2, precisely in the band number 2q13 and 2q14.1 and technically this specific location of fusion has been designated as 2qFus (Fan et al, 2002). The presence of head to head telomere repeats in chromosome number 2 of the human genome has been interpreted as the unique way of the emergence of chromosome 2 in the human genome, as an accidental fusion of telomere to telomere (Idol et al, 1991). It rendered the molecular biologists to contemplate that the fusion assumed to have occurred with duplication of DNA sequence in subtelomeric sequences present at least one of the chromosomal end of the fusion partners of the ancestral chromosomes (Fan et al, 2002).

In his well-defined article, entitled "*One pedigree we all may have come from- did Adam and Eve have the chromosome 2 fusion?*" on evolutionary emergence and its inheritance of chromosome 2 in hominins, Pawel Stankiewicz, Department of Molecular and Human Genetics, Baylor College of Medicine, Houston, USA, has come up with a hypothesis to define the occurrence of HSA_2, fusion as a "*non-recurrent rearrangement event which has substantial "evolutionary advantage due to genetic loss*"" (Stankiewicz, 2016). In his hypothesis, Stankeiwicz (2016) had proposed that instead of occurrence of recurrent Robertsonian translocation, which had genetically the most prevailing molecular chromosomal maneuvering, witnessed in the human genome, the fusion of HSA_2 as well as its emergence in hominins has been contemplated as an occurrence of a single non-recurrent event that transcended through a small-sized polygamous clan population bottleneck. Stankeiwicz (2016) has further defined uniquely as he said: "*Its heterozygous to homozygous conversion, fixation, and accumulation in the succeeding populations was likely facilitated by an evolutionary advantage through the genomic loss rather than deregulation of expression of the gene(s) flanking the HSA_2 fusion site of 2q13.*" During his molecular investigations on the HSA_2 fusion site on chromosome 2q13, Stankeiwicz (2016) had failed to recognize any HSA_2 junction polymorphisms in the modern human genome that rendered him not to recognize fusion of HSA_2

as a recurrent genomic event. It intrigued Stankeiwicz (2016) to come up with a proposition as he said: "*I propose that HSA_2 arose only once likely in one early modern human male and was subsequently transmitted and accumulated as a heterozygous event, converted to the homozygous state due to inbreeding in a small 'bottleneck' polygamous clan, and spread in the succeeding populations.*" *In the research communication, entitled "Nonhomologous Robertsonian translocations (NHRTs) and uniparental disomy (UPD) risk: an Italian multicentric prenatal survey,*" Italian research scientists Sensi and the fellow research associates have briefly narrated an evolutionary elucidation of the erstwhile population of modern human species of polygamous pedigree in which prehistoric implementation of HSA_2 had been reported followed by its molecular transformation from heterozygous state to homozygous state (Sensi et al, 2004). The findings of Sensi et al (2004), has been graphically defined by Stankeiwicz (2016) and he further contended an analogy, identical scenarios, happened to the carrier of Robertsonian translocations and expected that a male carrier of HSA_2 chromosome from 2nd generation, who supposedly engaged in a polygamous relationship with few female counterparts, supposedly yield multiple progenies in 3rd generation and out of the total progenies, 48 are supposed to be balanced healthy (with normal chromosomal profile) and 47 progenies with chromosomal imbalance, along with the substantial rate of miscarriages. The graphical elucidation of Stankeiwicz (2016) has further narrated that random mating among the individuals of 47 with a chromosomal profile of heterozygous for HSA_2 type likely to produce 48 balanced healthy, 47 healthy and 46 (having a chromosomal profile with homozygous for HSA_2) in IV the generation. Stankeiwicz (2016) has contemplated from such a presentation that the individuals, 46 with HSA_2 chromosomes in a homozygous state if ever need to go through in course of random inbreeding, its potential level of fertility would have remained unaffected, and inbreeding depression would not be reflected in its future population dynamics.

Stankiewicz (2016) has further extrapolated that the enhanced level of fertility, which has been triggered by the gene testis-expressed gene $DDX_{11}L_2$, might be regarded as an evolutionary advantage. In their meticulous investigations, the molecular geneticists and genome biologists have observed that the genes, found to be in the proximity of the fusion site of HSA_2, were noticed to be functionally restricted in site-specific action like gene expression of *$DDX_{11}L_2$* have been restricted to the testis, whereas gene expressions of *$RPL_{23}AP_7$* and *$CBWD_2$* have been restricted to brain region and the gene expressions of *$WASH_2P$*, *$FAM_{138}B$*, and *$RABL_2A$* have found to be equally functional in brain and testis, rather than any other sites of the body of modern human species (Stankiewicz, 2016). Scientists have further contended that most likely the change of transcription profile of the genes in brain and testis might help the modern species to be emerged with better cognitive potential and a higher level of fecundity in modern human species, possessing HSA_2 either in the homozygous or heterozygous state, in comparison to their ancestral lineage of apes (Stankiewicz, 2016). Though Stankiewicz (2016) had tried to give an alternative explanation by referring to the potential loss of some genomic material in HSA_2, once found to have existed in the identical location of ancestral chromosomes (IIp and IIq) of Great apes. By comparing a tentative occurrence, related to such evolutionary process, Stankeiwicz, (2016) has argued in favor of his second hypothetical explanation as he thought the evolution of the early humans likely to be expedited, mainly due to loss of genes in the subtelomeric regions of the ancestral chromosomes of apes, genetically recognized as IIp and IIq, respectively though there is a little gap in between fact and prediction here, though as the genome biologists are not currently aware of the gene content of this region of the Great apes. Several molecular investigations on the evolutionary effect of irreversible gene-loss on a different group of animals, specifically the primates, revealed that the act of pseudogenation and loss of genes had reported triggering the genetic variation in micro-level and rendered speciation of the concerned group of animals, that supposedly happened in speciation of primates, in course of its future evolution of hominins (Olson, 1999; Hahn et al, 2007). The hypothetical contemplation of "gene loss as a pervasive engine of evolution and speciation of primates" had further been ascertained through the brief reviewing of empirical comparisons of loss of genes in the genomes between hominin primates and non-human primates and the qualitative results revealed that the extent of loss of the gene in non-human primates was found to be considerably high in comparison to the human primates (Demuth et al, 2006; Wang et al, 2006).

Pertaining with the emergence of a higher level of cognitive-behavioral potential in course of human evolution and primate speciation, Stankeiwicz (2016) had further defined that the assertive mating (not random) between individuals characterized with a higher extent of cognitive potential, happened due to potential genomic loss of genes (either heterozygous or homozygous), found to be promoting structural and functional development of the brain and it would likely cause evolutionary divergence of HSA_2 in the next level. In support to his genealogical explanation of the evolution of hominins, through the evidence of HSA_2, Stankiewicz (2016) has firmly stated: "*I hypothesize that heterozygous and homozygous genomic loss of the putative gene(s) embedded in the repetitive subtelomeric sequences might have played a more important evolutionary role than deregulation of gene expression and/or a change of function of the genes flanking the* HSA_2 *fusion site.*"

The HSA_2 fusion site has genetically mapped the first time within the telomeric repetitive sequences, identical to the small arms of the orthologous chimpanzee chromosomes PTR_{12} and PTR $_{13}$ (Ijdo et al, 1991; Kasai et al, 2002). At least a couple of earnest endeavor has been made by Fan along with two separate groups of researchers to trace out the low copy repeat or LCR-rich genomic structure and gene content of ~600kb location, surrounding the fusing site nearby 2q13 and they mapped the chromosome junction in the degenerated telomeric arrays in an intron of the non-coding $DDX_{11}L_2$ transcript and the molecular biologists have further noticed that this fusion site is resistant to PCR amplification and cloning (Fan et al, 2002a; Fan et al, 2002b). The contemporary molecular investigations of LCR location in chimpanzee and human revealed that there is a 40kb genomic fragment, found in the four locations of the human genome, but it is 400 times within the chimpanzee genome (Cheng et al, 2005). The scientists further proposed that "*amplification of this interval occurred before fusion of* HSA_2" (Stankiewicz, 2016). The scientists have further observed that LCRs, adjacent to the ancestral fusion site, supposedly promoted the mix n match type of reorganization of closely identical telomeric repeats of the two ancestral chromosomes (Stankiewicz, 2016). In a contemporary molecular evolutionary hypothesis, promulgated by Ventura et al (2012), it has been contemplated that ancestral human-chimpanzee chromosomes, the occurrence of HSA_2 fusion and pericentric inversion of chromosomal maneuvering supposedly protected the chimpanzee and human genomes, to form subtelomeric heterochromatin. After going through a series of molecular investigation, like a review of chromosomal alignments about the active and vestigial centromeres, the contemporary gene profile of HSA_2,

PTR_{12} and PTR_{13} chromosomes, scientists have ascertained that a megabase of repetitive satellite-rich, subtelomeric DNA sequences had found to be missing from both of the ancestral chromosomes, recognized as IIp and IIq, respectively and the deletion of such DNA sequences supposedly happen during the emergence of the infusing chromosome, known as HSA_2 (Avarello et al, 1992; Ventura et al, 2012). One of the unique research communications, entitled "*Parallel patterns of evolution in the genomes and transcriptomes of humans and chimpanzees*," from Khaitovich and his research colleagues had critically analyzed the molecular footprints of around 21,000 protein-encoding genes of the transcriptomes, triggering the structures and functions of heart, kidney, brain, testis, etc. of human and chimpanzees, its nearest, extant, phylogenetically and the empirical results on this molecular investigation on the gene sequences and gene expression related to specific tissues between these two species have found to be identical (Khaitovich et al, 2005).

The contemporary molecular investigations, done by molecular geneticists and genomicists, the subtelomeric location of the human chromosome of the human genome was found to be one of the functionally dynamic spots as the DNA sequences of non-homologous chromosomes have recurrently been duplicated, exchanged, and recombined (Mefford and Trask, 2002). However, scientists have contemplated that such fusion involved loss and rearrangement of genetic materials evidenced by the absence of some genes, originally found in the two separated original chromosomes, possessed by the apes. In their computer-aided investigations on genomic comparisons of AT-GC base-pair substitutions between human species and chimpanzees, genomicists have used rhesus macaque as a reference and estimated that around 0.74 mya the fusion of ancestral chromosomes have been taken place, with a 95% confidence interval of 0 to 2.81 mya (Dreszer et al, 2007). The molecular investigations of our immediate line of extinct predecessors like Denisovans, Neanderthals, etc. have the identical existence of fused chromosome number 2 in their genome, as we carried it in our genome and it pointed out further that such accidental fusion of chromosomes in apes likely to be happening around 0.75 to 4.5 mya. (Meyer et al, 2012). On their research investigation, entitled "The evolution of African great Ape subtelomeric heterochromatin and the fusion of human chromosome 2," Ventura et al (2012) have calibrated the molecular clock to contemplate the tentative age of fusion of chromosome 2, that supposedly occurred > 4 mya in the evolutionary history of hominins. Very recently, genomicists have critically investigated the satellite DNA at pseudo-centromere at 2q21.2 in the HSA_2 (Avarello et al, 1992) and estimated that fusion of HSA_2 supposedly taken place around 3.5 mya (within a range of 2.5 to 4.5 mya) (Miga, 2019). However, an account of comparative genomic analysis among Neanderthals, Denisovans, and the modern Human species revealed that structurally and functionally the two extinct hominins (Neanderthals and Denisovans) and the extant modern humans have an identical chromosomal profile that rendered the inter-specific cross among all hominins yield viable (fertile) offsprings. As there was no viable trace cross in the nature between hominids (having 48 pairs of chromosomes) and hominins (46 pairs of chromosomes) and it has generally been observed that genetical incompatibility supposedly render no viable offspring, hence chromosomal fusion might have a suitable analogy of reproductive isolation that not only minimize the chance of emergence of interspecific offspring, it acts as an evolutionary shut off valve that minimizes evolutionary reversal of human species to apes (Rejon, January 17, 2017). It needs to be mentioned here that in 7th April 2005, detailed research analysis on chromosome number 2 and 4 of the human chromosome had been published in *Nature*, one of the most reputed journals, where the scientists have find out the that there are a large "gene deserts" known in Human genomes and it has further re-affirmed the erstwhile observation of molecular biologists that emergence of human chromosome number 2 has taken place as a result of 2 ancestral chromosomes of apes and it needs to be mentioned here that this research project was part of Human genome project in Washington, coordinated by a big crowd of research scientists from multiple research institutions and the project has been sponsored by NHGRI or National Human Genome Research Institute and National Institute of Health (Bethesda, April 6, 2005).

Every coin has two sides, and consideration of the validity of any evolutionary model or any hypothesis whether it's functional or not has often been challenged by an opposing group of evolutionists. The hypothetical reduction of chromosome count from 24 pairs to 23 pairs and the assumption that the reported fusion or reorganization, taken place in a particular site of chromosome 2, seems to be a simple assumption without any valid evidence as the assumed fusion sites of chromosome 2 contain some functional genes, that might not promote any such molecular amalgamation or genetic fusion, so the critiques of the evolutionary hypothesis of "Human Chromosome 2 Fusion" claimed it as a perceptional prediction rather than a scientific observation (Bergman and Tomkins, 2011). In support of the hypothesis of humans from apes, a contemporary group of evolutionists, recognized as Biologos community claimed from their meticulous observation on the sequential arrangement of DNA on the chromosomes of apes and humans and stick to their hypothesis to promote that "*There is no good biological reason to find the same genes in the same order in unrelated organisms, and every good reason to expect very different gene orders*" (Jeanson, September 14, 2019). But it has been strongly been criticized by the supporters of YEC, who opposed the ape to man hypothesis, as the contended that the above statement has not enough scientific merit in it as the experimental review of the functional analysis of genes among different group of primates, failed to comply the evolutionary assertion, supposed to be witnessed among apes and modern human species (Jeanson, September 14, 2019).

To judge any hypothetical proposition regarding the evolution of human species in the mindset of accepting reality, it's not necessary to be part of YEC supporter or to be a radical atheist of the Biologos community, rather it is the proper way to embrace the truth. Dr. Francis S. Collins, the Director of NHGRI, who was the in-charge of Human genome Project in 2005, has made an eloquent observation on the molecular investigation, specially made on Human chromosome number 2 and chromosome number 4 as he commented "*This analysis is an impressive achievement that will deepen our understanding of the human genome and speed the discovery of genes related to human health and disease. In addition, these findings provide exciting new insights into the structure and evolution of mammalian genomes*" (Bethesda, April 6, 2005). The meticulous investigations of the molecular biologists have ascertained that the number 2 human chromosomes are the second-largest chromosome (in length), followed by the Chromosome number 1 and the scientists had found that it is the treasure trove of the gene with the longest protein-coding sequence of 280,000 bp long gene, that encodes "Titin," a muscular protein, consists of 33,000 amino acids (Bethesda, April 6, 2005). In the contemporary molecular investigation, the scientists had engaged in finding out the DNA sequence of the "relic" centromere region of the ancestral ape chromosome, which has been inactivated after fusing with another ancestral ape chromosome, found to be possessing 36,000 base-base of DNA sequence, that precisely pointed out the inactive centromere region. The contemporary molecular investigation has ascertained further that this relic centromere tract has been earmarked with a certain type of DNA duplication, recognized as “alpha satellite repeats” and the trach has been flanked by another type of DNA duplication here, called “segmental duplication” (Bethesda, April 6, 2005). The main objective of the Human Genome Project (HGP) was to identify the functional stretches of DNA or gene to find out what kind of protein that it would encode for and the genomicists have noticed that human chromosome number 2 is the holy grail of 1,346 protein-encoding genes, whereas the chromosome number 4 is the treasure trove of 796 protein-encoding genes (Bethesda, April 6, 2005), so if the combination of deletion, alteration, duplication of some sequence happened at the time of accidental fusion of two ancestral chromosomes in apes, it steered the evolution of hominins to let emerge modern human species and rendered that it adapted physically, physiologically and immunologically to be diverged out as a unique species that could micromanage every ecosystem on Earth. As a part of the investigation of human chromosome number 4 of HGP, scientists have noticed that a vast array of chromosome number 4 has found to be devoid of any protein-encoding genes, recognized as “Gene Deserts,” the exact function of this part of the human genome is needed to be explored in future but the evolutionary biologists construed that such evolutionary hot spots would help to decipher human evolution in a better way as these hot spots have found to be conserved throughout the evolution of Mammalians and Avis (Bethesda, April 6, 2005). In a breakthrough in the field of molecular genetics and genomics, scientists have explored an mRNA transcript from a gene, belonging to chromosome 2 of modern human species and this particular gene, that produces a protein, found to be present in the genome of the Chimpanzee and the modern human species. Referring to the “tentative” evidence, it has been assumed that the gene had supposedly made a protein in the brain and testis in the human species and the researchers have located “hypervariable” region in the chromosome number 2, the genes stationed in such locations have found to be functionally variation type so, functionally accustomed to produce a various number of altered proteins, supposedly back-up evolutionary challenges of hominins, that human population face on its lone evolutionary journey of life on Earth (Bethesda, April 6, 2005).

Though the occurrence of chromosomal aberration is a rare incident, yet scientists have witnessed such a rare occurrence (O'Neill, 2010; Bi et al, 2012). The rare occurrence of "Robertsonian translocation" has been reported in 13 families so far in a homozygous state (44 chromosomes), which has been referred by the scientists (Neu et al, 1975; Rockman-Greenberg et al, 1982; Morgan et al, 1985; Abdalla et al, 2013; Song et al, 2016). Recently, a group of scientists have narrated that a healthy male 44, XY, der (14; 15) (q10; q10), der (14; 15) 9 (q10; q10) male, having no record of the irregular formation of sperm, belonging to a group of individuals (homozygous for a particular Robertsonian translocation chromosome), that sustained under long term isolation, might theoretically put forth a new human subspecies, genetically characterized with 44 chromosomes (Song et al, 2016). Though the occurrence of such new subspecies out of the root-stock of modern human species has remained a theoretical possibility. If we made a brief comparison of the socio-behavioral sexual orientation of the modern human species with its predecessors, we could easily notice that the contemporary modern human species are found to be very much accustomed to monogamous socio-behavioral orientation, in contrary, the erstwhile ancestors of the modern human species have been accustomed to polygamous socio-behavioral orientation, like the Big apes (e.g. Gorilla), used to engage with (Nelson et al, 2011). So, a group of evolutionists has contemplated that such a shift of socio-behavioral features and sexual orientation from polygamy to monogamy promoted (act as selection pressure) wide-spread dispersal of HSA_2 and further converted heterozygous state to homozygous state (Stankeiwicz, 2016). The evolutionary biologists, physical anthropologists, molecular geneticists, genome biologists have considered that emergence of HSA_2 in modern human species should inevitably be considered as an important evolutionary incident that not only initiates the evolutionary journey of hominins, diverged out of the ancestral root-stock of hominoids, it also helped the emergence of hominins with the unique appearance like upright profile, bipedal mode of walking, better cognitive functions like communicating abilities by using languages, better-memorizing potential, the high extent of running endurance, enhanced level of thermos-regulation by utilizing sweat glands, witnessed in different subspecies of hominins. The evolutionary biologists have construed that the backtracking investigation on the emergence of HSA_2 pertaining with immediate ancestral lineages of Homo sapiens like *Homo heidelbergensis* or *Homo erectus* should give an idea of the emergence of HSA_2 in terms of its evolutionary age and its molecular foot-prints in course of evolutionary divergence modern human species from the ancestral lineage of Great apes (Stankeiwicz, 2016). But the molecular biologists found that the main impediment for such investigation is the degradation of the genomic DNA of the ancestral, extinct members of hominins and its irreversible change of biochemical properties, the mutation cascades that usually burn the bridge to participate in its evolutionary roll-back its ancestral state. Several evolutionary biologists have expected further that successful reconstruction of "Great Apes orthologous subtelomeric chromosomes" would likely help a better understanding of its evolutionary role in the evolution of hominins through the formation of HSA_2. The formation of HSA_2 and the genes, resident of the proximity of its fusion site, that help to enlarge brain and better brain functions, the enhanced reproductive functions in testis and the switching mode of bio-behavioral, social and reproductive orientation from polygamy to monogamy seems to be a paradigm shift in the evolutionary journey of humans, that has been traversed through the crossroad of long-stretch of the evolutionary trail and perceptional contender of Young Earth Creationists (YEC), the ultimate symbolic labyrinth of Adam, Eves and Apes (Dumas et al, 2012; Stankeiwicz, 2016).

References

Abdalla, E. M., Kholeif, S. F., and Elshaffie, R. M. (2013). Homozygosity or a Robertsonian translocation (13q; 14q) in an otherwise healthy 44, XY man with a history of repeated fetal losses. *Lab Medicine*. 44: 254–7.

Avarello, R., Pedicini, A., Caiulo, A., Zuffardi, O., and Fraccaro, M. (1992). Evidence for an ancestral alphoid domain on the long arm of human chromosome 2. *Hum Genet*. 89(2):247-9. doi: 10.1007/BF00217134.

Baltimore, D. (2001). Our genome unveiled. *Nature*. 409:814–816.

Bethesda, M. (April 6, 2005). Scientists analyze chromosomes 2 and 4. NHGRI-supported researchers discover largest "Gene Deserts;" Find new clues to ancestral chromosome fusion event

https://www.genome.gov/13514624/ 2005-release-scientists-analyze-chromosomes-2-and-4

Bergman, J. and Tomkins, J. (2011). The chromosome 2 fusion model of human evolution—part 1: Re-evaluating the evidence. *Journal of Creation* 25(2):106–110.

Bi, W., Probst, F.J., Wiszniewska, J., Plunkett, K., Roney, E. K., Carter, B. S., et al. (2012). Co-occurrence of recurrent duplications of the DiGeorge syndrome region on both chromosome 22 homologues due to inherited and de novo events. *J Med Genet*. 49:681–8.

Cann, R.L., Stoneking, M. and Wilson, A.C. (1987). Mitochondrial DNA and human evolution. *Nature*. 325: 31-36.

Carter, R. and Hardy, C. (2015). Modelling biblical human population growth. *Journal of Creation* 29(1):72–79.

Cheng, Z., Ventura, M., She, X., Khaitovich, P., Graves, T., Osoegawa, K., Church, D., DeJong, P., Wilson, R.K., Pääbo, S., Rocchi, M., and Eichler, E. E. (2005). A genome-wide comparison of recent chimpanzee and human segmental duplications. *Nature*. 437(7055): 88-93. doi: 10.1038/nature04000.

Conrad, D.F., Keebler, J.E., DePristo, M.A., Lindsay, S.J., Zhang, Y., Casals, F., Idaghdour, Y., Hartl, C.L., Torroja, C., Garimella, K.V., Zilversmit, M., Cartwright, R., Rouleau, G.A., Daly, M., Stone, E.A., Hurles, M.E., Awadalla, P.; 1000 Genomes Project. (2011). Variation in genome-wide mutation rates within and between human families. *Nat Genet*. 43(7): 712-4. doi: 10.1038/ng.862.

Demuth, J.P., Bie, T.D., Stajich, J.E., Cristianini, N., and Hahn, M.W. (2006). The Evolution of Mammalian Gene Families. *PLoS ONE* 1(1): e85.

https://doi.org/10.1371/journal.pone.0000085

DeWitt, D.A. (January 14, 2014). What about similarity between human and chimp DNA?

https://answersingenesis.org/genetics/dna-similarities/what- about-the-similarity-between-human-and-chimp-dna/

Dreszer,T. R., Wall, G.D., Haussler, D., and Pollard, K.S. (2007). Biased clustered substitutions in the human genome: the footprints of male- driven biased gene conversion. *Genome Res.* 17(10):1420-30. doi:10.1101/gr. 6395807.

Dumas, L.J., O'Bleness, M.S., Davis, J.M., Dickens, C.M., Anderson, N., Keeney, J.G., Jackson, J., Sikela, M., Raznahan, A., Giedd, J., Rapoport, J., Nagamani, S.S., Erez, A., Brunetti-Pierri, N., Sugalski, R., Lupski, J.R., Fingerlin, T., Cheung, S.W., and Sikela, J. M. (2012). DUF1220- domain copy number implicated in human brain-size pathology and evolution. *Am J Hum Genet.* 91(3):444-54. doi: 10.1016/j.ajhg. 2012. 07.016.

Eck, A. (August 3, 2013). Tracking down our genetic "Adam and Eve."
https://www.pbs.org/wgbh/nova/article/our-genetic-adam-and-eve/

Fan, Y., Linardopoulou, E., Friedman, C., Williams, E. and B. J. Trask, B. J. (2002a). Genomic structure and evolution of the ancestral chromosome fusion site in 2q13□2q14.1 and paralogous regions on other human chromosomes. *Genome Research* 12(11):1651–1662.

Fan, Y., Newman, T., Linardopoulou, E., and Trask, B.J. (2002b). Gene content and function of the ancestral chromosome fusion site in human chromosome 2q13□2q14.1 and paralogous regions. *Genome Research* 12(11): 1663–1672.

Francalacci, P., Morelli, L., Angius, A., Berutti, R., Reinier, F., Atzeni, R., Pilu, R. et al. (2013). "Low-Pass DNA Sequencing Of 1200 Sardinians Reconstructs European Y-Chromosome Phylogeny." *Science* 341(6145): 565–569.

Grehan, J.R. and Schwartz, J.H. (2009). Evolution of the second orangutan: phylogeny and biogeography of hominid origins. *Journal of Biogeography* 36(10): 1823-1844.

Hahn, M.W., Han, M.V., Han, S-G. (2007). Gene Family Evolution across 12 *Drosophila* Genomes . *PLoS Genet* 3(11): e197.
https://doi. org/10. 1371/journal.pgen.0030197

Hardy, C., and R. Carter. (2014). The biblical minimum and maximum age of the earth. *Journal of Creation* 28, no. 2: 89–96.

IJdo, J.W., Baldini, A., Ward, D.C., Reeders, S.T., and Wells, R.A. (1991). Origin of human chromosome 2: an ancestral telomere-telomere fusion. *PNAS U S A.* 88(20):9051-5. doi: 10.1073/pnas.88.20.9051.

Jeanson, N.T. (2017). *Replacing Darwin: The new origin of species*. Master Books. Kindle Edition. File Size 11575 kb.

Jeanson, N.T. (2019). Testing the Predictions of the Young-Earth Y Chromosome Molecular Clock: Population Growth Curves Confirm the Recent Origin of Human Y Chromosome Differences.

https://answersingenesis. org/theory- of-evolution/molecular-clock/testing-predictions-human-y- chromosome- molecular-clock/

Jeanson, N.T. (September 14, 2019). Genetics confirms the recent, supernatural creation of Adam and Eve. Featured in "*Searching for Adam.*"

https://answersingenesis.org/adam-and-eve/genetics-confirms-recent-supernatural-creation-adam-and-eve/

Jeanson, N. T. (April 25, 2021). The best-kept secrets of DNA.

https:// answersingenesis.org/genetics/best-kept-secrets-dna/

Kasai, Y., Kishira, H., and Harayama, S. (2002). Bacteria belonging to the genus Cycloclasticus play a primary role in the degradation of aromatic hydrocarbons released in a marine environment. Appl. Environ. Microbiol. 68: 5625-5633.

Khaitovich, P., Hellmann, I., Enard, W., Nowick, K., Leinweber, M., Franz, H., Weiss. G., Lachmann, M., and Pääbo, S. (2005). Parallel patterns of evolution in the genomes and transcriptomes of humans and chimpanzees. *Science*. 309(5742): 1850-4. doi: 10.1126/science.1108296.

Langergraber, K.E., Prüfer, K., Rowney, C., Boesch, C., Crockford, C., Fawcett, K., Inoue, E., Inoue-Muruyama, M., Mitani, J.C., Muller, M.N., Robbins, M.M., Schubert, G., Stoinski, T.S., Viola, B., Watts, D., Wittig, R.M., Wrangham, R.W., Zuberbühler, K., Pääbo, S., and Vigilant, L. Generation times in wild chimpanzees and gorillas suggest earlier divergence times in great ape and human evolution. *PNAS U S A*. 109(39): 15716-21. doi: 10.1073/pnas.1211740109.

Mefford, H. C., and Trask, B. J. (2002). The complex structure and dynamic evolution of human subtelomeres. *Nat Rev Genet*. 3(2):91-102. doi: 10.1038/nrg727.

Meyer, M., Kircher, M., Gansauge, M.T., Li, H., Racimo, F., Mallick, S., Schraiber, J.G., Jay, F., Prüfer, K., de Filippo, C., Sudmant, P.H., Alkan, C., Fu, Q., Do, R., Rohland, N., Tandon, A., Siebauer, M., Green, R.E., Bryc, K., Briggs, A.W., Stenzel, U., Dabney, J., Shendure, J., Kitzman, J., Hammer, M.F., Shunkov, M.V., Derevianko, A.P., Patterson, N., Andrés, A.M., Eichler, E.E., Slatkin, M., Reich, D., Kelso, J., and Pääbo, S. (2012). A high-coverage genome sequence from an archaic Denisovan individual. *Science*. 338(6104):222-6. doi: 10.1126/science.1224344.

Miga, K. H. (2019). Centromeric Satellite DNAs: Hidden Sequence Variation in the Human Population. Genes (Basel). 10(5):352. doi: 10.3390/genes10050 352.

Mitchell, M.W., Locatelli, S., Ghobrial, L. *et al.* (2015). The population genetics of wild chimpanzees in Cameroon and Nigeria suggests a positive role for selection in the evolution of chimpanzee subspecies. *BMC Evol Biol* 15, 3 (2015). https://doi.org/10.1186/s12862-014-0276-y

Morgan, R., Bixenman, H., and Hecht, F. (1985). Human chromosome variation with two Robertsonian translocations. *Hum Genet*. 69:178–80.

Mortenson, T. (2014). *Searching for Adam: Genesis & the truth about man's origin.* Master Books. Leaf Publishing group, USA. P.524.

Nelson, E., Rolian, C., Cashmore, L., and Shultz, S. (2011). Digit ratios predict polygyny in early apes, Ardipithecus, Neanderthals and early modern humans but not in Australopithecus. *Proc Biol Sci.* 278(1711): 1556-63. doi: 10.1098/rspb.2010.1740.

Neu, R. L., Valentine, F. A., and Gardner, L. I. Segregation of a t (14q22q) chromosome in a large kindred. *Clin Genet.* 8: 30–6.

Noonan, J. P. (2010). Neanderthal genomics and the evolution of modern humans. *Genome Res.* 20(5): 547-53. doi: 10.1101/gr.076000.108.

Olson, M.V.(1999). When less is more: gene loss as an engine of evolutionary change. *Am J Hum Genet.* 64(1):18-23. doi: 10.1086/302219.

O'Neill, I. D. (2010). Homozygosity for constitutional chromosomal rearrangements: a systematic review with reference to origin, ascertainment and phenotype. *J Hum Genet.* 55(9):559-64. doi: 10.1038/jhg.2010.80.

Purdom, G., Jeanson, N.T., and Mortenson, T. (March 1, 2019). Making the leap from Ape to Adam.

https://answersingenesis.org/human-evolution/making- leap-ape-adam/

Rejon, M.R. (January 17, 2017). The Origin of the Human Species: a Chromosome Fusion?

https://www.bbvaopenmind.com/en/science/ bioscience/the-origin-of-the-human-species-a-chromosome-fusion/

Rockman-Greenberg, C., Ray, M., Evans, J. A., Canning, N., and Hamerton, J. L.(1982). Homozygous Robertsonian translocations in a fetus with 44 chromosomes.*Hum Genet.* 61:181–4.

Rohde, D. L., Olson, S., and Chang, J.T. (2004). Modelling the recent common ancestry of all living humans. *Nature.* 431(7008):562-6. doi: 10.1038/ nature02842.

Sensi, A., Cavani, S., Villa, N., Pomponi, M.G., Fogli, A., Gualandi, F., Grasso, M., Sala, E., Pietrobono, R., Baldinotti, F., Savin, E., Ferlini, A., Cecconi, M., Rossi, S., Gallone, S., Bellini, C., Neri, G., Martinoli, E., Simi, P., Dalprà, L., Genuardi, M., Dagna-Bricarelli, F., and Calzolari, E.(2004). Nonhomologous Robertsonian translocations (NHRTs) and uniparental disomy (UPD) risk: an Italian multicentric prenatal survey. *Prenat Diagn.* 24(8):647-52. doi: 10.1002/pd.962.

Song, J., Li, X., Sun, L., Xu, S., Liu, N., Yao, Y., Liu, Z., Wang, W., Rong, H., and Wang, B. (2016). A family with Robertsonian translocation: a potential mechanism of speciation in humans. *Mol Cytogenet.* 9: 48. doi: 10.1186/s13039-016-0255-7.

Stankiewicz, P. (2016) One pedigree we all may have come from – did Adam and Eve have the chromosome 2 fusion? *Mol. Cytogenet.* 26: 72.

The 1000 Genomes Project Consortium. (2015) A global reference for human genetic variation. *Nature* 526: 68–74. https://doi.org/10.1038/nature15393

The Chimpanzee Sequencing and Analysis Consortium. (2005). Initial sequence of the chimpanzee genome and comparison with the human genome. *Nature* 437: 69–87. https://doi.org/10.1038/nature04072

Tijo, J.H. and Levan, A. (1956). The chromosome number in Man. *Hereditas* 42(1- 2): 1-6.

Trasancos, S. (March 28, 2017). How do Adam and Eve fit with evolution?

https://www.ncregister.com/blog/how-do-adam-and-eve-fit-with- evolution

Tomkens, J.P. (December 28, 2011). Genome-Wide DNA Alignment Similarity (Identity) for 40,000 Chimpanzee DNA Sequences Queried against the Human Genome is 86–89%.

https://answersingenesis.org/genetics/dna- similarities/genome-wide-dna-alignment-similarity-identity-for-40000- chimpanzees/

Venn, O., Turner, I., Mathieson, I., de Groot, N., Bontrop, R., and McVean, G. (2014). Nonhuman genetics. Strong male bias drives germline mutation in chimpanzees. *Science*. 344(6189): 1272-5. doi: 10.1126/science.344. 6189.1272.

Ventura, M. , Catacchio, C.R., Sajjadian, S., Vives, L., Sudmant, P. H., T. Marques-Bonet, T., Graves, T.A., Wilson, R.K., and Eichler, E.E. (2012). The evolution of African great ape subtelomeric heterochromatin and the fusion of human chromosome 2. *Genome Research* 22(6):1036–1049.

Wang, X., Grus, W.E.,and Zhang, J. (2006). Gene Losses during Human Origins. *PLoS Biol* 4(3): e52. https://doi.org/10.1371/journal.pbio.0040052

Wong, K. (September 1, 2014). Tiny Genetic Differences between Humans and Other Primates Pervade the Genome.

https://www.scientificamerican. com/article/tiny-genetic-differences-between-humans-and-other-primates- pervade-the-genome/

Chapter 5

Probability of a Bottleneck of a Primordial first pair at Origin of Humanity on Earth, around 500,000 Years Ago: A Hypothetical Appraisal

At the moment, the Ayala paper appears to be the only study which shows more than 4 allele lineages with trans-species variation. His analysis, however, did not estimate confidence nor did it use phylogenetics to determine lineages (it used sequence similarity instead). In 22 years, I cannot find a paper that replicates his finding with better methods (or any methods). Certainly, trans-species variation has been observed, but not more than 4 lineages, as far as I can tell. This is not enough evidence by which to make a confident claim against a single generation bottleneck.

---- *Joshua Swamidass*

"*Darwin's Radio*," won prestigious awards in the literary world like the Nebula award, and Endeavour award (2000), which was considered to be one of the most eloquently written Science Fiction novels in the last decade (Lynch, 2001), authored by Greg Bear, who put together a state of art biological and anthropological research to accommodate the contemporary advancement of molecular genetics and genomics, virology and molecular evolution to narrate the origin and evolutionary divergence of humans and their civilization, that might supposedly reach at the scary, cross-road of biological evolution in unforeseeable future. Bob Kurland, a retired physicist from old school orientation, and a committed follower of Christian faith has referred to this classic Science fiction novel as a unique literary work that has feasibly addressed the theological issues, when most science fiction did not get close to it (Kurland, March 3, 2019). The story in this science fiction novel revolves around its central character Kaye Lang, an eminent molecular biologist from the United States, a specialist in the endogenous retrovirus, a predominant RNA virus that invaded with its nucleic material into the host's DNA and finally integrated to the host's genome (Bear, 2000). However, the story started with a business trip of Kaye Lang to Georgia, one of the independents of the erstwhile Soviet Union, to re-establish the business relationship and scientific cooperation among the scientists and expert virologists in that area with other parts. On a side trip, she engaged herself in a scientific investigation of the exhumed bodies of the pregnant women of a mass grave in Georgia, and she got in touch with virus hunter Christopher Dickens at the Center for Disease Control (Epidemic Intelligence Control) and briefly narrated and effectively pursued the trail of a flu-like disease, called "Herod's flu," caused by Human Endogenous Retrovirus or HERV, that attacked the young would be mothers, leading to the abortion of their foetus before birth. After going back to US, Kaylee Lang, came to know that Herod's flu is nothing but a dreadful incident of "Scattered Human Endogenous retro-Virus Activation" (SHEVA). The scariest part of the story had been associated with the ulucidation of the genetic information of SHEVA, that was found to have been laying dormant for a long period of time, approximately millions of years, had woken up to lead to the outbreak of an epidemic, that rendered all pregnant women to miscarry. In a twist to this story, an important character, anthropologist, Mitch Rafelson had been introduced, who investigates three 15,000 year old, mummified bodies of pre-historic human bodies located in a mountain cave in the Alps, out of three, two matured individuals were recognized as one male and one female Neanderthals, both of them HERV, that caused outbreak of SHEBA epidemic and the infant had been found to be the modern human species. The research observations made by Mitch had intrigued him to contemplate the evolution of the modern human species out of the Neanderthal parental lineages and get him in touch with Lang and Dicken to establish his discovery with the findings of activation of SHEBA and the Georgian mass grave, to reconstruct a zig-Shaw puzzle of the historical origin of past, present and future (which is merely predictive) evolutionary diversification of modern human species on Earth. Greg's evolutionary hypothesis of human species out of Neanderthal's lineage is quite intriguing as the genome biologists have noticed in their investigation that various human lineages of modern human species possess around 3% of Neanderthal genes (Kurland, December 19, 2015).

Digging deeper, down the enthralling, adventurous trail of cutting-edge inventions of molecular biology and contemporary principles of evolution, Darwin's Radio had presented a brief narrative of activation of endogenous retrovirus, SHEVA in its fast-paced thriller as well as one of the most unique science fictions in this century (Bear, 2000). Here in this science fiction, the evolutionary journey of SHEVA has been depicted in a unique way, where it was contemplated to evolve to its next generation and to integrate to the evolutionary mechanism of humans, to initiate further programming of speciation, when inside the uterus of the pregnant woman. The contemporary progress in genome biology, specifically in the Human Genome Project (HGP), has elucidated that a large part of the human genomes are occupied by non-coding DNAs, or introns and some parts of the introns have found to be remnants of retroviruses, those triggered to be activated, accidentally start translating a number of "Large Protein Complexes" or LPC. Greg Bear has tried to postulate activation and speciation, triggered by an intricate genetic network, possessed by the host organism (here it's the human body), intrigued by the adaptive response to overcrowding in human population (seems to be auto-thinning out evolution diversification process in the history of humanity). Similar to the transmission of sexually contracted disease, SHEVA was contemplated to be contracted by woman through unprotected lateral transmission from their male counterparts and existed there in a living state for a period of time. If the women got impregnated, the foetus would have reportedly miscarried, a deformed female fetus in its first trimester (third month of pregnancy) left behind a fertilized egg with 52 chromosomes, paving the way for emergence of newly diverged future or unique form of derived species of *Homo sapiens sapiens*, out of the root-stock of modern human species possessing typically, 46 number of chromosomes, instead. The second stage of pregnancy in the third trimester would be contemplated to be born with higher IQs than modern human species with novel characteristics, such as an altered form of sense organs and sensitivity to the viral infections (to be viable or born), if their mothers were subjected or exposed to such diseases like measles or chicken pox etc. and the usage of anaesthetic to the pregnant women at the time of child-birth has related to giving birth to stillborn. In order to avert such a disastrous emergence of SHEVA-induced divergence of the unique human species, two hypothetical options have been proposed, the first one was very much straight forward, recommending the abortion of the SHEVA infected second stage foetus to inhibit its appearance or inhibition of

SHEVA gene activation by identifying the source of its signal of activation and block the molecular pathway, but it's difficult to do as the genetic material of Herod's flu virus is present in the genome of the contemporary human race and due the high pace of mutation of the "missing-link signal molecule," that trigger the activation of the SHEVA gene, the adoption of feasible inactivation measure of SHEVA inflicted emergence of novel or next generation human species has been narrated a difficult task. Though the entire story of Darwin's Radio has presented a great story of origin and evolution of humanity, and its civilization in the backdrop of the molecular biological advancement of science, a number of supported of liberal creationists like Bob Kurland believed that the author Greg Bear had addressed the key concern about the origin of human life and its emergence in the perspective of creationism as it had been addressed by Greg Bear "*What is the mechanism by which the creator might form the new species*?" (Kurland, March 3, 2019). In his well-written elucidation, titled "***Were Adam and Eve Real? Evolutionary Bottlenecks and "Darwin's Radio***,"" Greg Bear had tried to raise a hypothetical concern about whether his speculative evolutionary mechanism that rendered to form a new species, integrated into the occurrence of an epidemic, caused by endogenous retrovirus, possessed a junk part of DNA, which was found to be part of modern human genome as well, is theoretically sound or not (Kurland, March 3, 2019). It has further raised another question in this novel "*who sends the signal for Darwin's Radio to sound?*" and which drags the evolutionary cause and effect type of interpretation of activation of the HERV epidemic to a theological twist, as it raised concerns that after a certain period of time lying dormant (hibernation), whether it was the hand of God or the "time span set by the designer nature" that designed (reset the evolutionary button) or redesigned the occurrence of mutations for paving the emergence of a species (Kurland, December 19, 2015). Greg bear has taken his speculative narratives of speciation, specifically the origin and evolution of the human species and its tentative future derived form to the next level perception as he said "*what you seem to be implying, however, is a hitherto undiscovered mechanism whereby the genome takes control of its own evolution, somehow sensing the right time to bring about change*" (Bear, 2000). We wish Greg Bear's speculative mechanism of human speciation is an evolutionary narrative rather than a speculative narration of science fiction that could define the origin and evolution of the modern human species and elucidate further the integrated issues like whether its emergence had taken place out of the primordial couple (Adam and Eve) or emerged out of any prehistoric population of human beings having more than two, whether the predecessor of the modern human species had ample extent of genetic diversities or not, and whether the modern human race had gone through any evolutionary bottlenecks of two individuals or not.

The diverse range of perception of theistic evolutionism itself do not promote any specific scientific principle or set of theories but it does help to accommodate a diverse range of scientific theories and hypotheses, explaining the functionalities of biological evolution to correlate to the religious beliefs and historical perceptions, rather than uphold the traditional beliefs of special creationism and it has been witnessed by the heterogeneous community of contemporary modern human race (comprised of subsect of the communities of science and the subsect of progressive minded individuals, those belonging to different religious communities) that the theistic evolutionists have presented a number of genetic and molecular analysis, arguments, and interpretations in favor of and against the possible existence of Adam and Eve as the rootstock of humanity in prehistoric time. The scientific probability of prehistoric existence of Adam and Eve has been strongly advocated in the book of "*Adam and the Ape*," authored by R.J. Berry in 1975 and he has put his claim on the basis of frequencies of variants of the protein haemoglobin, and he hypothesized that around millions of years ago the human species was comprised of a single pair for a single generation or an erstwhile population of 10,000 for 500,000 years or somewhere in between 2 and 10,000 (Berry, 1975). Though Berry's (1975) observation on the variants to consider them as "selectively neutral" was found to be partly true, it failed to substantiate strongly that humanity has transcended out of the ancestral root-stock of Adam and Eve in prehistoric times (Buggs, April 14, 2018). In his article, titled "*Polymorphism in mitochondrial DNA of humans as revealed by restriction endonuclease analysis*," which has been published in the 77th volume of *Proceedings of National Academy of Science, USA* in 1980, Wesley M. Brown has critically analyzed the variation of mitochondrial DNA of the human species and explored "*the amount of sequence heterogeneity observed, 0.18%, could have been generated from a single mating pair that existed 180-360 x 10^3 years ago, suggesting the possibility that present-day humans evolved from a small mitochondrially monomorphic* population that existed at that time"

(Brown, 1980). In the article, entitled "Mitochondrial DNA and two perspectives on evolutionary genetics," which was published in 1985 in the 26th volume of "Biological Journal of Linnaean Society," Allan C. Wilson, along with a number of research associates have reviewed a comparative studies of animal mitochondrial DNA (mtDNA) and put together two distinct lines of progress in evolutionary biology. One line dealt with studies of genealogical trees, time scales and new mutations on surviving molecular lineages and the other lineage dealt with determining population structure, migration and frequencies of alleles, found in ancestral population. The critical observation of Wilson and his colleagues (1985) have helped to contemplate that humanity has traversed through a short, sharp bottleneck, as the molecular investigation of the ancestral human population revealed that the variability of mtDNA has found to be anomalously low though a normal level of nuclear variability has been witnessed in its genome (Wilson et al, 1985). Following the footsteps of Allan Wilson, in 1987, Rebecca Cann, along with Mark Stoneking and Allan Wilson (posthumous) have published their work in "Nature," in which it has been claimed that mitochondrial DNA the modern human race had a common ancestry in Africa around 200,000 years ago (Cann et al, 1987). Cann and her research colleagues had engaged in measuring the mitochondrial differences among major racial groups of modern human species. Based on an assumptive, constant rate of change of evolution, Cann and her colleagues had used a common mitochondrial sequence and extrapolated backtracking to a most recent common ancestor. The ancestor has logically been recognized as mitochondrial Eve by the team Cann, as mitochondria have been found to be inherited through maternal lineage and it has also popularly been recognized as "African Eve hypothesis" (Cann et al, 1987). An interesting research article has been published in the 268th volume of the journal "Science," entitled "*Absence of polymorphism at the ZFY locus on the human Y chromosome*" which has been authored by R.L. Dorit, H. Akashi and W. Gilbert in 1995 and it published the result of the molecular investigation of the particular segment (this segment of Y chromosome is comprised of 729 base pairs) of Y chromosomes, randomly collected from 38 men, representing the different ethnic groups all over the world. Surprisingly, the researchers did not find any variations in their analysis and that rendered them to contemplate that modern human race must have gone through the genetic bottleneck in their recent past (Dorit et al, 1995). Further confirmatory observations made the investigators declare that any living male of modern human species, descended from a single male or a very small group of males, scientifically named as "Y-Chromosomal Adam" who lived around 270,000 years ago (Dorit et al, 1995; Wilford, May 26,1995). It needs to be mentioned here that John Noble Wilford, Science correspondent of New York Times had made an interesting story in 26th May, 1995, on origin and evolution of the modern human species, titled "*Believers in African Eve think they have found Adam(s)*" mainly based on the research publications Cann et al (1987) and Dorit et al (1995), which got a wide media coverage at that time. As Y-chromosome has been inherited by the male offspring from their father (paternal lineage), the mitochondrial DNA descends from mother to the daughter only (Maternal lineage) and the scientists have further revealed that all living individuals, belonging to modern human species either male or female, descended out of a single woman or certain female ancestors lived in Africa around 200,000 years ago, recognized as "Mitochondrial Eve" (Dorit et al, 1995; Wilford, May 26, 1995). In their investigations, Dorit et al (1995) have further ascertained that Y-chromosomal Adam was supposedly the sole survivor before he had children but it was not the case for the Mitochondrial Eve (not considered to be only women before having children), rather she has scientifically been recognized as one of the ancestor of all living children and this (Wilford, May 26, 1995).

Most of the theological supporters of Abrahamic religions have followed a moderate line of the perception of origin and evolution of human species as they accept the principles of evolution and they believe that the modern human race have descended from a single couple and it has been considered valid by a number of evolutionists and a large part of communities belonging to science. But a review of such perceptions was found to be necessary along with a publication of a scientific/non-fiction, entitled "*Adam and the Genome: Reading Scriptute after Genetic Science*," written by Dennis R.Venema, along with Scot McKnight the eminent Christian Biologist from Trinity Western University, British Columbia, Canada in 2017. In his literature, "*Adam and the Genome: Reading Scripture after Genetic Science*," Venema has cited three references of population genetical studies, that were able to determine that the size of the population of the erstwhile predecessor of the modern human race has restricted to be around 10,000 individuals, so the single–pair bottleneck in the history of humankind has been considered to be next to impossible in the contemporary findings of mainstream science (Buggs, October 28, 2017). Venema, has stated his theoretical perception in order to negate the remote possibility of evolutionary bottleneck of a pair of two as he said " *If a species were formed through such an event [by a single ancestral breeding pair] or if a species were reduced in numbers to a single breeding pair at some point in its history, it would leave a telltale mark on its genome that would persist for hundred years- a severe reduction in genetic variability for the species as a whole*" (Buggs, October 28, 2017). In support of his point of arguments, Venema has cited the example of the substantial loss of genetic variability of the Tasmanian Devil, a carnivorous marsupial species found to be in pandemic in distribution all over Australia but found to be endemic in the Tasmanian coast of Australia. The population biologists have observed that the genetic variability of the Tasmanian Devils is reported to have been lost in last few hundred years and most of the individuals have the same alleles in their genome, with little differences and it has also been analyzed by Venema that spreading out of a type of transmissible, facial cancer in the population of Tasmanian Devil communities has found to be attributed in lowering its genetic diversity (Buggs, October 28, 2017). Hence, the reference of evolutionary bottleneck of Tasmanian Devils helps to hypothesize that if the modern human population had ever gone through a bottleneck, it would have the identical instance of depletion of genetic erosion. But an opposing group of population biologists have brought a new twist in it by defining it distinctly. Scientifically, it is a matter a of fact that Tasmanian Devils have gone through a bottleneck but the nature of distinct type bottlenecks varies in the length and intensity of it and its consequence on different species, and in most occasions, scientists have observed that impact of long bottleneck was found to be more severe on the same species than it exposed or sustained to short bottleneck (Buggs, October 28, 2017). The population biologists have further hypothesized that a sharp and short-term bottleneck exposure is likely to have happened in order to amalgamate the historical story of Adam and Eve and to integrate the biological perception of evolution of the human species.

The observation and analysis of Venema (2017) was critically challenged by Richard Buggs, the evolutionary genomicist from Queen Mary University of London who has refuted Venema's observation by sending a personal e-mail in the month of May, 2017, where Buggs claimed that the fundamental rules of population genetics reveal that surging a great deal of genetic diversity in a short period of time, from a small population (even from a founding population of two) is possible (Evolution News, January 25, 2018). In "*Adam and the Genome*," Venema has theorized that high level of heterozygosity could sustain the bottleneck, but high level of allelic diversity could not sustain it. So, it is convincing that the contemporary findings of high level allelic diversity in the modern human species would not fit the hypothesis that this species has traversed through a bottleneck of two in prehistoric time. Refuting Venema's theoretical explanation, Buggs (November 12, 2017) has responded with his online article or blog: http:// richardbuggs. com/ response_to_ Dennis_ Venema_ Biologos_Partl_Adam_Eve. html) to Venema, where Buggs has expressed his concern about the reference of the impact of population bottleneck on Tasmanian devils, who was once found to be wide-spread in Australia but found to be restricted in some island or coastal regions as he mentioned "*...got an impression you are including heterozygosity within your definition of genetic variability because of your choice of Tasmanian Devils as an exemplar of a species that has undergone a bottleneck. This exemplar takes up quite a large proportion of the early part of your chapter. It is well known that Tasmanian devils have low heterozygosity as well as low allelic diversity- they have much lower levels of heterozygosity than humans...*" (Evolution News, January 25, 2018).

The publication of Venema'a book has revived the old debate in a new form at Biologos which continued for a period of time about whether we have a historical origin and legacy of evolution from a bottleneck of two primordial couple or likely to be descended from an ape-like population of more than several thousands. The debate has been triggered by the comments, made by Richard Buggs, Reader in Evolutionary genomics at Queen Mary University of London, UK when he criticized the narratives of Venema, mentioned in his book, titled "*Adam and the Genome: Reading Scripture after Genetic Science*." In support of his arguments, Venema (2017), has stated "*DNA evidence indicates that humans descend from a large population because we, as a species, are so genetically diverse in the present day that a large ancestral population is needed to transmit that diversity to us. To date, every genetic analysis estimating ancestral population sizes has agreed that we descend from a population of thousands, not a single ancestral couple. Even though many of these methods are independent of each other, as methods employed to date agree that the human lineage has not dipped below several thousand individuals for the last three million years or more- long before our lineage was even remotely close to what we would call 'human.' Thus the hypothesis that humans descend solely from one ancestral couple has not yet found any experimental support- and it is therefore not one that genetics view as viable*" (Gauger, March 5, 2018). Dr. Venema has found to be confident in contending that the research findings of population biology and molecular genetics did not acknowledge the traditional claim of the historic existence of Adam and Eve as the early couple that had initiated erstwhile journey of humanity on Earth. As he expressed his firm assertion that his observation on the earliest population size of the modern human race (which has been contemplated to be around 10,000) is considered to be scientifically consistent like "Heliocentric model of the solar system," which has been mentioned in his book, titled "Adam and the Gene" as he stated "*The Sun is at the center of our solar system, humans evolved, and we evolved as a population.*" *Put most simply, DNA evidence indicates that humans descend from a large population because we, as a piece, are so genetically diverse in the present day that a large ancestral populations needed to transmit that diversity to us. To date, every genetic analysis estimating ancestral population sizes has agreed that we descend from a population of thousands, not a single ancestral couple* (Venema, 2017).

Buggs has tried his best to sort out the perceptional differences between him and Venema earlier in the month of May, 2017, regarding the difference of opinion contended by him and Venema on the origin and evolution of the modern human race on Earth. But he did not get any response to his personal email communications sent to Venema, and after waiting for month and he decided to share it in public portal (www.richardbuggs.com/email-to-dennis-venema-adam-eve-genomics.html), where Buggs (September 29, 2017) had raised a number of concerns about the scientific feasibility of Venema's observation and interpretation about the evolutionary beginnings of humanity on earth; as Buggs has stated "*If population growth is faster than this [become double or more in every generation], the proportion of the heterozygosity maintained will be higher. This means the pair of individuals can carry a great deal of heterozygosity with them through a bottleneck, if they come from an ancestral population with high diversity, and they will pass that on to the population they found, so long as it grows rapidly.*"

Buggs (September 28, 2017) further stretched out his logical arguments, which has literally refuted Venema's argument in his personal e-mail (www. richardbuggs. com/email-to-dennis-venema-adam-eve-genomics.html) as he stated "*As you will know, it is a feature of humans that despite our current census population size of over seven billion individuals, we have lower genetic diversity than the world's much smaller current day population of chimpanzees. The average human has 3.1 million single nucleotide variants (SNVs), but the average chimp has 5.7 million (Prado-Martinez et al, 2013 Nature). African humans approximately 1.1 heterozygous SNVs in every 1000bp, whereas central chimpanzees have approximately 1.75 (Prado-Martinez et al 2013 Nature). Thus, if two central African chimpanzees were taken today and used to found an isolated population that experienced explosive population growth, the new population would have similar of genetic variability to modern humans. I am not stating these figures because existing populations of chimpanzee gave rise to modern humans, but simply to show that it is hard to see how overall levels of SNP diversity and heterozygosity in modern humans could exclude the possibility of a past bottleneck of two individuals. On the top of this, we need to add the fact that explosive population growth in humans has allowed many new mutations to rapidly accumulate in human populations, accounting for many SNPs with low minor allele frequencies (A. Keinan and A.G. Clark, 2012, Science).*"

However, the perception of a short-term bottleneck and its impacts are found to vary from person to person among the population biologists. As Ernst Mayr (1963), one of the renowned scientists in molecular and evolutionary biology and population biology, had observed that the species had gone through extreme evolutionary bottlenecks in founder effects, the extreme loss of the genetic diversity of such species, would put them through the trail of evolutionary changes (Buggs, October 28, 2017). On the contrary to the observation of Mayr (1963), the other population biologists of the opposing group have appeared with a counter hypothesis which claimed that even an acute bottleneck of a single pair of individuals would not abruptly diminish the genetic diversity if it offset by an immediate and rapid population growth (Nei, Maruyama and Chakraborty, 1975). Experimentally, it has been observed by scientists that if two individuals have been sampled out of a fairly large population of a biological species, on average, each individual possess 75% of its heterozygosity and if a derived line of populations formed out of the bottleneck of a single fertilized egg and the population doubles in every generation, the level of heterozygosity would be bounced back to more than half of the level of that derived line of population in reference to its previous level of heterozygosity, maintained prior to bottleneck event (Barton and Charlesworth, 1984; Slaktin and Excoffier, 2012). So, it is convincible that replenishment or bouncing back of the heterozygosity level of any derived population in the post-bottleneck regime depends on the reciprocal rate of population growth of that species, hence, it is most likely that the Tasmanian Devil has sustained a sharp and short stint of evolutionary bottleneck exposure, without any back-up of exponential population growth that rendered its present scanty size of population which was restricted in distribution in some coastal / island regions of Australia, that might not be the situation that the human race had went through, even it would have reportedly gone through the short and sharp stint of bottleneck.

It needs to be mentioned here though Dennis Venema and Richard Buggs both are aligned to distinct perceptions to interpret the origin and evolution of modern human race, both reached in a consensus that understanding of functionalities of heterozygosity and allelic diversity in defining bottleneck. But Richard Buggs has accused Venema regarding improper usage of the term "genetic diversity" as he elaborated it in his email communication to Venema, which has been published on 12th November, 2017, in his online article or blog (http://richardbuggs.com/response_to_Dennis_Venema_Biologos_Partl_Adam_Eve.html) as he explained: "*...in scientific population genetics literature the term "genetic diversity" does not refer only to allelic diversity. Genetic variability of populations is measured in many ways: heterozygosity, allelic diversity, private allele frequency, gene diversity, fixation indices, inbreeding coefficients etc. I did not realise that when you use the term in your chapter you intend only to refer to allelic diversity. That is not the way the term is normally used in the field. I therefore assumed that you were also referring to heterozygosity. It is a pity that this ambiguity was present, but I understand that it is hard to write about science at a popular level without the occasional ambiguity slipping in that a specialist will stumble on*" (Evolution News, January 25, 2018).

The population biologists have noticed that bottleneck has diminished the allelic richness rather than bringing down the rate of heterozygosity of any species (Leberg, 1992). In reference to the diminishing level of allelic richness in relation to the bottleneck and got an assumption about the size of a population, Venema has stated "*...scientists have many other methods at their disposal to measure just how large our population has been over time. One simple way is to select a few genes and measure how many alleles of that gene are present in present-day humans. Now that human genome project has been completed and we have sequenced the DNA of thousands of humans, this sort of study can be done simply using a computer. Taking into account the human mutation rate, and the mathematical probability of new mutations spreading in a population or being lost., these methods indicate an ancestral starting point of just two individuals, one who have to postulate mutation rates far in excess of what we observe for any animal*" (Buggs, October 28, 2017). In absence of any direct reference, substantiated by Venema and McKnight (2017), Buggs (October 28, 2017) has extrapolated by giving a feasible explanation as he has mentioned that a there were four number of alleles that can go through a bottle-neck of two individuals and in regards to the Single nucleotide polymorphisms (SNPs), the number of alleles at SNP is four (A or Adenine, T or Thymine, G or Guanine and C or Cytosine. The Human Genome Project has ascertained that the common SNPs in humans is reported to be biallelic. Buggs (October 28, 2017) has further referred that consideration of contiguous sequence of DNA, like a Haplotype of a gene, where would witness more than four alleles, those are likely acquired new mutations or combined through ancestral SNPs (Keinan and Clark, 2012).

Venema and McKnight (2017) has referred to the work of Tenesa et al (2007) to review the effective population size (Ne), estimated from "Linkage disequilibrium," which stated further that human population size has hardly been traversed through a bottleneck of two and Venema has further mentioned "*The results indicate that we come from an ancestral population of about 10,000 individuals-...obtained when using allele diversity alone.....scaled its analysis to investigate our lineage from the present dating back to approximately 200,000 years ago....when our species first appears in the fossil record [Venema does not consider Homo erectus to be human]. The researchers found that, during this period, humans living in sub-Saharan Africa maintained a minimum population of about 7,000 individual, and value other methods arrive at*" (Buggs, October 28, 2017). As the observation of Ne size of less than 10,000 individuals, that has been mentioned in the work of Tenesa et al (2007), has not supported the

Venema's calculative determination of the size of Ne sizes by combining the population size of above and below the Sahara and it further clarified that non-sub-Saharan populations possesses "*subset of the amount of genetic variation present in the African population*" and it has supported the "Out–of–Africa" hypothesis and historically as well as scientifically ascertained the expansion of prehistoric ancestral population of sub-Saharan region and finally results revealed that the size of the "Ne" of sub-Saharan region is around 6282 with a standard deviation of 1357 and this study has been determined based on the determination of recombination rate of population size (Buggs, October 28, 2017).

Alternatively, the population biologists have used Linkage disequilibrium method to determine the effective size of a population, given a recombination rate and adoption of this methodologies by the scientists has brought down the previous figure of Ne by 27%, and which has considered to be feasible worldwide (Buggs, October 28, 2017). It is matter of fact that nowhere is it clearly mentioned in any investigations that the size of the effective population has been narrowed down to two on the basis of some substantial evidence, but it does not disprove effectively either as the experimental model, Tenesa and his fellow researchers have adopted in 2007, has clearly defined that that there has been an assumption that the population of given time points are considered to be in an equilibrium state rather than considered to be in exponential population growth but in reality, exponential growth of human population that has taken place in the last 10,000 years, that has been categorically overlooked so the feasibility of determination of effective size of population less than 6000 number is still open. Venema and McKnight (2017) has referred the work of Li and Durbin (2011) that has adopted "*Pairwise Sequential Markovian Coalescent Model*" (PSMC) to analyze the singe genome to determine the size of the effective population from the analysis of the heterozygosity, witnessed in the human genome. So, the effective size of the population went down to 5,700 from African individuals, which apparently does not show that the chance of getting closer to the bottleneck of two to consider it as the root-stock of humanity. However a number of population biologists have argued that a single couple can carry 0.75 of the heterozygosity of their ancestral population, so PSMC (as a model not so precise) would not effectively detect the short and sharp bottleneck caused by the sudden outbreak of global epidemics or natural disaster. So far, it is apparently premature to discount a perception of bottleneck to rift open the bottleneck issues in the arena of conflict between science and theology, but it is the matter of fact that as long as the findings of Venema and McKnight (2017) is concerned we did not have enough evidence to substantiate undoubtedly, that we the modern human race could have descended from the evolutionary bottleneck of prehistoric, primordial couple.

However, two contemporary experts in molecular biology specifically in molecular genetics and genome biology, scientists Steve Schaffner of the Board Institute of Cambridge, Massachusetts and Joshua Swamidass, assistant professor of the Department of Pathology and Immunology at Washington State University, St. Louis, have re-entered the debate, left behind by Venema and Buggs. Dr. Schaffner has carried out a computer-run simulation model to analyze SNP allele frequency spectra (AFS) from the African population in the 1000 genomes project to figure out whether a bottleneck of two individuals was empirically possible or not and he contended that around 500,000 years ago, the possibility of a bottleneck could not experimentally ruled out (Gauger, March 5, 2018; Buggs, April 14, 2018). Schaffner's analysis of allele frequencies had not distinctively differentiated between the allele frequencies analyzed at the end of the simulation of the bottleneck of two and presented by the contemporary genetic data. Whereas Joshua Swamidass has estimated the tentative time of the most recent four alleles possessed by the human genome. In a simple word, if allele is a gene, we carry two copies of each gene and a bottleneck of two individuals would possess four alleles per gene. So, as per molecular backtracking investigations of Swamidass, it has been observed that a bottleneck of recent time of two individuals or four alleles supposedly occurred around 500,000 years ago (Gauger, March 5 2018). It has been ascertained that both investigators have validated the mutation rates in their research models. So, the work of Schaffner and Swamidass had squared it out by technically blocking the possibility of emergence of the modern human species from a prehistoric population of a few thousand humans and opened up the possibility of a bottleneck of two as well as the emergence of a primordial couple (Adam and Eve) as the ancestral rootstock of journey of humanity on Earth. In an alternate way, the works of Sawmidass and Shaffner could also accommodate the emergence of modern human race out of apes lineage too as scientists would not rule out the bottleneck happened between 500,000 years ago and 7 mya, whenever the modern human species

(*Homo sapiens sapiens*) split out of its phylogenetic cousin the chimpanzee (*Pan troglodytes*) (Gauger, March 5, 2018). Though the technical findings of Swamidass and Schaffner had refuted the firm affirmation, made by the earlier population biologists, that the modern human race emerged out of a prehistoric population of a few thousand individuals and it has indicated a probability of existence of a primordial maiden couple on Earth, though it could not ascertain its existence but it stopped the denial of a possibility that the earlier scientists claimed that a bottleneck of two had never happened, though it has remained in the argumentative stage in Biologos, which requires more clarification and needs to be in form of a peer-reviewed valid publication to come out as a ground-breaking revelation in the contemporary history of human evolution.

Molecular geneticists have identified the functionalities of genes with multiples alleles as a real challenge against the hypothesis of the possible bottleneck of two in relation to origin and evolution of human on Earth. During execution of the Human Genome Project (2005), the genome biologists have noticed that there are human genes having as many as thousand alleles and those gene are functionally integrated to the immune system to protect the body against disease and parasites. The population biologists have contemplated further that the extent of diversity of such genes, possessed by the maximum individuals of a population, of the concerned population would have been attributed to sustain unforeseeable challenges of survival in the regime of rapid environmental changes. However, the molecular genomicists have isolated a number of human leucocyte antigen (HLA) alleles, found to be present in the identical genes in Chimpanzees and Spider monkeys (macaques) and it facilitate scientists to construe that the common predecessor of these three descendants had supposedly possessed and passed it on to its successors at the time of divergence of these three successors Gauger, March 5, 2018). So, theoretically, the evolutionary biologists have contended that each species inherited a subset of alleles from their ancestors at the time of their divergence and overlapping of these subsets happened between the species, sharing these alleles. The molecular geneticists have contemplated further "*that the subset of alleles inherited by each species would have been too numerous to have come through a population of bottleneck of two*" and it is hypothetically recognized as "trans-species polymorphism" or TSP (Gauger, March 5, 2018). Literally, TSPs are genetic variance, the origin of which predates their occurrence of speciation, ended in the sharing of alleles among phylogenetically related species and it was posed as the main issues in the argument against the probability of a small, sharp bottleneck (Klein, 1987). Jan Klein (1987) in his research article, "*Origin of major histocompatibility complex polymorphism: The trans-species hypothesis*" and Francisco J. Ayala along with three research colleagues (1994) in their research article "*Molecular genetics of speciation and human origins*" have strongly supported the TSP hypothesis and contended HLA genes of the immune system as main example of TSPs (Gauger, March 5, 2018). According to the hypothetical narrative of biological evolution, the two identical genetic sequence, apparently resembled each other but was not found to be related by having a common descent and had emerged by means of "convergent evolution." So, according to this hypothesis, alleles resembling each other are supposed to be emerged with identical sequence but evolved independently, rather than considered to be related by a common descent. After examining the extensive allelic diversity of the "major histocompatibility complex" or MHC gene DRB1, Ayala and the fellow researchers construed "*that human ancestral populations could never have been smaller than two or three thousand individuals at any time over last several million years*" (Ayala et al, 1994). Based on their research findings, Ayala and his colleagues have discarded the possibility of bottleneck of two and the "Adam and Eve" hypothesis as well. In 1996, there was an article entitled "*The Evolution of Human Populations: A Molecular Respective*" having been published fifth volume of the journal "Molecular Phylogenetics and Evolution," jointly written by Francisco J. Ayala and Ananias J. Escalante, who contended that human evolution has gone through the trail of repeated speciation and series of intermittent morphological changes from Australopithecus to *Homo habilis* to *Homo erectus* and *Homo sapiens* and their hominoid ancestors to orangutans, gorilla, chimpanzee and humans. Ayala and Escalante (1996) have observed that MHC consists of a number of polymorphic gene loci and they have critically studied and shared their observation that the genealogy of 19 human alleles of DQB1 locus of MHC gene coalesces more than 30 mya, prior to the divergence of old world monkeys and most of the human alleles have found to be genetically identical to pongid (any member of Great Apes) and cercopithecoid rather than other human alleles. Using the theory of gene coalescence Ayala and Escalante (1996) have predicted "*...these polymorphisms require human* *populations of the order of N=100,000 individuals for the last several million years*." Based on their investigation on mtDNA, Y-chromosome, and microsatellite autosomal polymorphisms and the consistency of MHC, Ayala and Escalante (1996) have contended that there was no sign of a narrow population bottleneck (like bottleneck of two) fewer than few thousands had ever taken place in the history of the origin of humanity. Furthermore, the analysis of computer simulation along with molecular data analysis had indicated that African origin of modern human species have been initiated with its migration "out of Africa," supposedly had taken place between 100,000 to 200,000 years ago (Ayala and Escalante, 1996).

The meticulous works of Ayala and his colleagues (1994) have intrigued Joshua Swamidass, Assistant Professor in Genomic Medicine Division, Washington State University, St. Louis, USA to initiate a logical criticism on disproving the existence of Adam and Eve and so-called reject of "occasional population bottleneck hypothesis" and made his observation available in the online article / blog: https://discourse.peacefulscience. org/t/ heliocentric-certainty-against-a-bottleneck-of-two/61, entitled "*Heliocentric certainty against a bottleneck of two*" and he stated "*...to be the only study which shows more than 4 allele lineages with trans-species variation. His analysis, however, did not estimate confidence nor did it use phylogenetics to determine lineages 9it used sequence similarity instead). In 22 years, I cannot find a paper that replicates his findings with better methods (or any methods). Certainly, trans-species variation has been observed, but not more than 4 lineages, as far as I can tell. This is not enough evidence by which to make claim against a single generation bottleneck*" (Swamidass, 2017, 2019). The molecular investigation of Swamidass had intended to find out if TSP should be considered as an ancient lineage that has gone through multiple events of speciation or it has been emerged as a result of convergent evolution. Like a number of genes, those involved in immune response, HLA genes, that complied the narrative of hypothetical emergence through convergent evolution has intrigued Swamidass to state about functionalities of HLA gene in his blog / online article: https://discourse. Peaceful science.org/t/heliocentric-certainty-against-a-bottleneck-of-two/61/18, and he has stated "*HLA-DRB1 is the most variable gene. It is notable for having over 500 squares in the DNA of about merely 1,000 individuals, compared with an expected number of less than 10. That means if we had we had tried to put the DNA into a tree, we would see at least 500 mutations discordant* *with a phylogenetic tree. This is just a stunning result, because it means that HLA-DRB1 alleles are just not well described tree. The variation we see is evolving and re-evolving over and over again.*" (Swamidass, February 5, 2018). Ann Gauger, a meticulous research investigator and postdoctoral fellow of Harvard University in developmental biology, who has authored "*Science and Human Origins*" along with two co-authors: Douglas Axe and Casey Luskin, in 2012, clearly elaborated the polymorphism issues, created by HLA-DRB1, one of the polymorphic gene of HLA and TSP in relation to the origin and evolution of the human species, criticizing the validity of undirected natural selection as primary evolutionary impetus of origin of the modern human race, out of the ancestral lineage of apes and they tried to afloat through critical reviewing of molecular data and analysis of molecular genetics and genomics whether there is any scientific probability to consider with whether the journey of humanity has started with the primordial couple on Earth or not? (Gauger et al, 2012). The critical arguments of Gauger et al (2012) in "*Science and Human Origins*" have clearly refuted the arguments of Francisco Ayala on HLA-DRB1 that denounced the possible existence of a first pair of couple as the root-stock of humanity on Earth, rater Gauger's observation , which was based one critical analysis of the evidence that revealed that trans-species polymorphism or TSP (which has considered to be happen as a result of convergent evolution) would not negate the probability of a bottleneck of two and it has revived the possibility of an occurrence of a bottleneck, around 500,000 years ago or earlier. It has further been substantiated from the molecular investigation of Joshua Swamidass, who has examined the genome-wide patterns of multiple haplotypes and put all findings together to find it out whether a bottleneck of two is plausible or not in the last few million years in the history of humanity and Swamidass had estimated the time to the most recent four lineages (TMR4L) for every recombination block in autosomal genome and his analysis indicated further probability of two-person bottleneck, likely around 500,000 years ago (Buggs, April 14, 2018). According to Richard Buggs, the adoption of advance level of molecular approaches like AFS and TMR4L seems to be feasible analysis of the data on human genetic diversity to validate Adam and Eve bottleneck but introgression[4] of Neanderthals and

---

[4] Introgression: The process, by which genetic principle of an species has been transferred to another species to yield a hybrid species as a result of repeated cycles of backcrossing.

Denisovans in the genome of modern human species, made the situation critical as Buggs theorize that the theory of a two person bottleneck seem to be plausible and tentatively occurred prior the divergence of modern human species on Earth, if Neanderthals and Denisovans had been considered within human lineage (Buggs, April 14, 2018). But Buggs (April 14, 2018) had expressed his concern about the alternative probabilities and he expressed "*If, on the other hand, we define only modern humans as truly "human," and posit a bottleneck within the human lineage, with later admixture of DNA variation from "non-humans," then a more recent bottleneck is possible, and the AFS and TMR4L approaches will be made very difficult to apply, unless admixed alleles can be filtered out.*"

References

Ayala, F.J., and Escalante, A. (1996). The evolution of human populations: A molecular perspective. *Molecular Phylogenetics and Evolution* 5(1): 188-201. DOI: 10.1006/mpev.1996.0013

Ayala, F.J., Escalante, A., O'Huigin, C. and Klein, J. (1994). Molecular genetics of speciation and human origins. *Proceedings of the National Academy of Sciences of USA* 91 (15):6787-6794; DOI: 10.1073/pnas.91.15.6787.

Barton, N., and Charlesworth, B. (1984). Genetic Revolutions, Founder Effects, and Speciation. *Annual Review of Ecology and Systematics*, *15*: 133-164.

Bear, G. [1999] (2000). *Darwin's Radio*. P.544. Ballantine Books.

Berry, R.J. (1975). *Adam and the Ape*. P. 80. Church Pastoral Aid Society.

Brown W. M. (1980). Polymorphism in mitochondrial DNA of humans as revealed by restriction endonuclease analysis. *Proceedings of the National Academy of Sciences of the United States of America*, *77*(6): 3605–3609. https://doi.org/10.1073/pnas.77.6.3605

Buggs, R. (18th April, 2018). Adam and Eve: Lessons learned.

http:// richardbuggs.com/index.php/2018/04/18/adam-and-eve-lessons-learned/

Buggs, R. (29th September, 2017). Email to Dennis Venema about human population bottlenecks.

http://richardbuggs.com/index.php/2017 /09/29/ email-to-dennis-venema-about-human-population-bottlenecks/

Buggs, R. (28th October, 2017). Adam and Eve: A tested hypothesis?

https://natureecoevocommunity.nature.com/users/24561-richard-buggs/posts/22075-adam- and-eve-a-tested-hypothesis

Buggs, R. (12th November, 2017). Response to Dennis Venema's Blog "Adam, Eve and Population Genetics: A Reply to Dr. Richard Buggs (Part 1)."

http://richardbuggs.com/index.php/2017/11/12/response-to-dennis-venemas-blog-adam-eve-and-population-genetics-a-reply-to-dr-richard-buggs-part-1/

Cann, R., Stoneking, M. and Wilson, A. (1987). Mitochondrial DNA and human evolution. *Nature* 325: 31–36 https://doi.org/10.1038/325031a0

Droit, R.L., Akashi, H. and Gilbert, W. (1995). Absence of polymorphism at the ZFY locus on the human Y chromosome. *Science* 268 (5214): 1183- 1185. DOI: 10.1126/science.7761836

Evolution News (25th January, 2018).

https://evolutionnews.org /2018/01/ adam-and-the-genome-and-human-genetic-diversity/

Gauger, A. (5th March, 2018). Is there a first human couple in our past? New evidence and arguments. http://evolutionnews.org/author/agauger/

Gauger, A., Axe, D. and Luskin, C. (2012). *Science and Human Origins*. P. 124. Discovery Institute.

Keinan, A., and Clark, A. G. (2012). Recent explosive human population growth has resulted in an excess of rare genetic variants. *Science 336*(6082): 740–743. https://doi.org/10.1126/science.1217283

Klein, J. (1987). Origin of major histocompatibility complex polymorphism: the trans-species hypothesis. *Hum Immunol.* 19(3): 155–162.

Kurland, B. (3rd march, 2019). Were Adam and Eve Real? Evolutionary Bottlenecks and "Darwin's Radio."

https://www.catholicstand.com/adam-eve-real- evolutionary-bottlenecks-darwins-radio/

Leberg, P.L. (1992). Effects of population bottlenecks on genetic diversity as measured by allozyme electrophoresis. Evolution 46(2):

https: //doi. org/ 10.1111/j.1558-5646.1992.tb02053.x

Li, H., and Durbin, R. (2011). Inference of human population history from individual whole-genome sequences. *Nature*, *475*(7357): 493–496.

https://doi.org/10.1038/nature10231

Nei, M., Maruyama, T., and Chakraborty, R. (1975). The Bottleneck Effect and Genetic Variability in Populations. *Evolution*, *29*(1): 1-10. doi:10. 2307/ 2407137

Prado-Martinez, J., Sudmant, P. H., Kidd, J. M., Li, H., Kelley, J. L., Lorente-Galdos, B., Veeramah, K. R., Woerner, A. E., O'Connor, T. D., Santpere, G., Cagan, A., Theunert, C., Casals, F., Laayouni, H., Munch, K., Hobolth, A., Halager, A. E., Malig, M., Hernandez-Rodriguez, J., Hernando-Herraez, I., … Marques-Bonet, T. (2013). Great ape genetic diversity and population history. *Nature*, *499*(7459): 471–475. https://doi.org/10.1038/nature12228

Slatkin, M., and Excoffier, L. (2012). Serial founder effects during range expansion: a spatial analog of genetic drift. *Genetics*. 191:171–181. URL http://www.ncbi.nlm.nih.gov/pubmed/22367031.

Swamidass, J. (30th December, 2017). Heliocentric certainty against a bottleneck of two? *Do "Humans" arise from a single couple?*

https://discourse.peacefulscience.org/t/heliocentric-certainty-against-a-bottleneck-of-two/61

Swamidass, J. (5th February, 2018). Heliocentric certainty against a bottleneck of two? *Convergent Evolution or Trans-Species variation?*

https://discourse.peacefulscience.org/t/heliocentric-certainty-against-a-bottleneck-of-two/61

Tenesa, A., Navarro, P., Hayes, B. J., Duffy, D. L., Clarke, G. M., Goddard, M. E., and Visscher, P. M. (2007). Recent human effective population size estimated from linkage disequilibrium. *Genome research*, *17*(4): 520–526.

https://doi.org/10.1101/gr.6023607

Venema, D. R. and McKnight, S. (2017). *Adam and the Genome*. P. 240. Brazos Press.

Wilford, J.N. (26th May, 1995). Believers in African Eve Think They have found Adam(s).

https://www.nytimes.com/1995/05/26/us/believers-in-african-eve-think-they-have-found-adam-s.html

Wilson, A.C., Cann, R.C., Carr, S. M., George, M., Gyllensten, U.B., Helm-Bychowski, K.M., Higuchi, R.G., Palumbi, S.R., Prager, E.M., Sage, R.D., and Stoneking, M. (1985). Mitochondrial DNA and two perspectives on evolutionary genetics, *Biological Journal of the Linnean Society*. 26(4): 375–400, https://doi.org/10.1111/j.1095-8312.1985.tb02048.x

Chapter 6

A Journey through Independent Evolutionary History of "*Mitochondrial Eve*" and "*Y-Chromosomal MRCA Adam*" on Planet Earth: A Temporal Snapshot

How can our male and female ancestors have lived thousands of years apart? Different genes have different history. One gene may sweep very quickly through an entire species, while another one takes much longer to spread. ………..The researchers [Scientists at the University of Arizona] collected DNA from men belonging to three population three populations- 25 Khoisan from Southern Africa, 24 Khalkhas from Mongolia, and 24 highland Papua New Guineans. Their ancestors branched off from one another tens of thousands of years ago. The results they found were surprisingly consistent: the women who bequeathed each set of men their mitochondrial DNA was twice as old as the man whose Y chromosome they shared. But the ages of Adam and Eve were different depending on which group of men the scientists studied. The Khoisan Adam lived 74,000 years ago, and Khoisan Eve lived 176,500 years ago. But the Mongolian and New Guinean ancestors were both much younger- Adam averaged 48,000 years old and Eve 93,000 years.

---- Carl Zimmer

According to the biblical narrative, the journey of humanity, which has been contemplated to have started from the primordial couple of Adam and Eve, had been accepted by theologists and member of churches, but it has emerged to be a perceptional hot potato, after the emergence of the Darwinian doctrine of evolution. The controversy has got a new lease of life with the emergence of molecular anthropological evidence of a primordial single woman, known as "mitochondrial eve" and the evidence of a primordial single man, known as "Y-chromosomal Adam," having come to the surface, and the contemporary analysis of the DNA molecules have rendered the evolutionists to construe that this primordial couple supposedly lived around 100,000 to 200,000 YBP, tracing back to be much earlier than the biblical timescale of the emergence of Adam and Eve in the Garden of Eden (Cann et al, 1987; Karafet et al, 2008). The availability of contemporary elaborative evidence of Y chromosomal and mitochondrial chromosome sequences and its analysis possibly yield a clear resolution of such evolutionary conflicts regarding the emergence of humanity on Earth. The molecular anthropologists have analyzed the data to understand it and interpret it in the output format of the historical changes in these two chromosomal entities starts with an exploration of DNA sequences of an extant population of modern human species, and gradual backtracking of its ancestral sequences of the archaic family groups, it belonged to in historic time. The initial endeavor of the analyses of the mitochondrial chromosome and Y chromosome have yielding distinct results that rendered the scientists to narrate the evolutionary history of humanity difficultly but it has found to be identical and overlapping in the contemporary investigations. In the field of evolutionary genetics, particularly in phylogenetics, and genealogical investigations of anatomically modern human species related to its origin and evolution, the utilization of mitochondrial DNA and Y-Chromosomal DNA has found to yield a tentative time of humanity's origin, broadly considered to be between 259,000 to 350,000 years ago, much earlier than the common perception, that claimed it supposedly evolved less than 100,000 years ago (Karmin et al, 2015). Several Young Earth Creationists (YECs) and contemporary evolutionists have referred the recent analytical investigations on Y chromosomal family tree and disagreed upon the earlier timescale for human origins, rather claimed that the origin of Adam and Eve, taken place around 6,000 years ago, led to the divergence of a most recent common ancestor (MRCA)

of the anatomically modern human race, around 4.500 years ago, at the time of the global flood (Hardy and Carter, 2014). Eventually, the perceptional rift between these two distinct models of the origin of humanity also recognized the history of events related to the human origin and its erstwhile size of the population, that impacted the scientific credibility of the biblical narrative of the origin of humanity on Earth. So, it gave an impression that in defining the origin and evolution of humanity, YEC and conventional evolutionists had taken its position at two opposite poles, though both have agreed to reconcile on the certain issues on the contemporary history of the origin of humanity, for last 3000 years pertaining with human events and population size and it is mainly based on the archaeological investigations to the biblical evidence (Jeanson, 2019). To substantiate such a claim, the researchers had tendered the archaeological recognition of the reign of King David around 1000BC (Coogan, 1998). The most unique findings to the researchers on this issue are that the biblical depiction of the temporal existence of King David is found to reconcile archaeological recognition (Hardy and Carter, 2014). But researchers carefully noticed that beyond 1000 BC, the postulations of the evolutionists and the YEC groups have failed to coalesce, in back-tracking the evolutionary history of humanity on Earth. To elucidate it further, researchers tendered the example of the biblical narrative of the exodus, that depicts the liberation of Israelites from the bondage (or slavery) from Egyptian and their historical journey to the Mount Sinai to form a new settlement, under the guidance of Moses and their passage through the Sea of Reeds, supposedly taking place around 1300 BCE. With little modification of the exact period, YEC research groups recognized the exodus of Israelites had taken place between 1400 to 1500 BCE, whereas the mainstream archaeological researches had not recognized the occurrence of such exodus and it intrigued researchers to construe that biblical narrative humanity had failed to reconcile the mainstream archaeological findings beyond 1000 BCE and that left a perceptional rift open between research groups, comprised of archaeologists, evolutionists, and inter-disciplinary and YEC researchers to give a feasible explanation of the emergence of humanity on Earth.

In resolving the controversy of determining the evolutionary age of mitochondrial Eve and Y-chromosomal Adam, the evolutionary geneticists and the interdisciplinary scientists like evolutionary biologists, molecular phylogeneticists, molecular anthropologists, etc., got involved and it indirectly prolonged the effort of reconciliation of scientific evidence, hypotheses and its interpretations to the biblical narrative of the history of the origin of humanity on Earth. Fazale Rana, one of the active participants in this process of reconciliation, who referred in his one of his well-circulated literature, titled "*Who Was Adam*?" that its molecular anthropologists, who played a cardinal role in deciphering the origins and evolution of modern humans on earth, where the genetic variability of the people all over the world has been critically analyzed (Rana, April 4, 2013).

The main intention of considering the DNA molecule to solve this puzzle of evolutionary mystery is primarily lying in its inherent property that acts as a blueprint of every human individual. But the most unique matter of DNA, is that the blueprint of DNA is often going through alteration in course of mutation and the mutation-induced variants of DNA molecule have usually passed on from one generation to the next generation. If inheritance of genetic variants (mutagen induced) descended from one generation to the next generation, and after several generations, we could expect a growing population, that is supposed to be comprised of individuals having a substantial accumulation of genetic variations that is a preponderance of genetic variants (Rana, April 4, 2013). The new population or the derived populations that predominantly represented by diversified genetic variants, found to be genetically distinct from its ancestral population, it diverged from, and these derived or break-way populations have been recognized as subpopulations by the population and evolutionary biologists. The anthropologists observed that the process of invasion, migration of the human population around the world causing the fragmentation or breakaway of the ancestral population, and divergence of several subpopulations. Eventually, these subpopulations with genetic variability have emerged with its own and distinct traits, and the evolutionary journey of the ancestral human population, ended up to a new beginning of the distinct evolutionary lineages belonging to anatomically modern human species.

Molecular anthropologists have chosen to focus their studies on the elucidation of the history of humanity to scientifically recording this act of divergence transcended from ancestral population to the derived one by "cataloging the genetic variants," based on the identical and distinct profile of DNA profiles among the intra and inter-population level of modern human species, found to be dispersed widely all over the world (Rana, April 4, 2013). These techniques, adopted by the molecular anthropologists to trace back all derived populations, coalesced from a common ancestral group as the scientists carried out an appraisal of the divergence of the diversified DNA molecules, and the tentative evolutionary age of the ancestral population of human beings and this popular technique has scientifically recognized as "molecular clock analysis." According to Rana (April 4, 2013), it seems to be analogical to an automobile, that travels to a distance in a period and if we know the speed of the automobile, the time of travels to certain destination could easily be calculated (Rana, 4th April; 2013):

Time = Distance / Speed

Likewise, if we're familiar with the numerical values of differences of two distinct genetic materials of two distinct populations are known and the rate of mutation is known then the tentative time of divergence between the two distinct populations could be calculated as follows (Rana, April 4, 2013):

Time = Number of genetic differences/rate of mutation.

To elucidate it further, Rana (April 4, 2013) has given a hypothetical reference of two distinct populations, the DNA molecules have been extracted from these two populations and have analyzed further, to notice that those two derived populations diverged out of the common ancestral population and found to be different in 100 locations and if the rate of mutation of the DNA molecule per location in each year is 1, those two distinct populations split out from one another around (100/1) = 100 years ago.

Based on their critical investigations, the molecular anthropologists have tried to interpret their experimental observation on the mitochondrial DNA variants, that not only helped them to trace out humanity's maternal lineage and the Y chromosomal DNA variants, that helped the scientists to trace out the paternal lineage of humanity but rendered them to contemplate that anatomically modern human species have emerged out of the primordial couple Adam and Eve on Earth (Rana, April 4, 2013). A further extrapolation of such observation, made by the molecular anthropologists helped the YEC researchers to stand behind the reconciliation of biblical narrative of the creation of humanity, that interpretation put the evolutionary perception of humanity on a stiff challenge. In support of their hypothesis the YEC researchers and a group of molecular anthropologists have claimed that the journey of the modern human race had been initiated from a small archaic human population, supposedly lived in a single location of East Africa and that intrigued the theologists to reconcile the biblical narrative of the "Garden of Eden" on Earth (Rana, April 4, 2013). Though the hypothetical interpretation of the emergence of humanity has been challenged and redefined by the next level of investigations, initiated by molecular anthropologists over the last two decades.

Though there is a valid reason of doubt to justify mitochondrial eve as the universal mother of modern human species, yet the symbolic motherhood of mitochondrial eve has been assumed to be biblical eve, the age dating endeavor of this motherhood initiated in 2005 and the evolutionary backtracking has claimed that mitochondrial eve supposedly lived in between 250,000 to 150,000 years ago (Rana, 14th April 2013). In this effort of molecular anthropology, the scientists engaged in comparison of relic genetic profiles to its derived ones by a quantitative appraisal of the rate of mutations of its inherent DNA molecule (Endicott et al, 2009). In 2009, a critical investigation on the comparative genetic account between chimpanzees and modern human species and further determination of the rate of mutation in their respective mitochondrial DNA, made by the molecular evolutionists to claim that they supposedly diverged out of a common ancestor around 6 mya (Endicott et al, 2009). Later the molecular geneticists have observed the rate of mutation is not constant, and it varied from different locations of the same genome and the identical organism possessing having an identical genome but lived in different geographical regions. However, the revised estimation of the evolutionary age of mitochondrial Eve, done by Scally and Durbin (2012) claimed to be between 300,000 to 250,000 years ago and they further observed that the rate of mutation of mitochondrial DNA, assumed to be little slow that rendered pulling back the evolutionary age of modern humans a little further down than previous observation. The recent comprehensive mitochondrial genomic studies of the 53 people around the world have helped the molecular anthropologists to determine the evolutionary age of the mitochondrial Eve, which supposedly emerged around 171,000+/-50,000 years ago (Rana, 14th April 2013). The inconsistent record of the evolutionary age of mitochondrial eve has been evidence in the different investigations, made by the different molecular anthropologists, as the rate of mutation has found to be less consistent and varies from one experimental model to another as the rate of mutations varies in the genomes of the populations of different generations. So, the scientists have adopted a unique methodology to undertake fine calibration of the mitochondrial DNA clock, by analyzing the DNA molecule, extracted from 10 human fossils, found to be 40,000 years old (determined by carbon dating process or C-14 dating process, used for archaeological investigation) (Fu et al, 2013). The consideration of archaeological period incorporating to the DNA analysis helped the investigators to undertake a fine calibration of the mitochondrial clock to put a feasible claim of the evolutionary age of mitochondrial Eve, which has found to be emerged around 157,000 +/- 40,000 years ago (Fu et al, 2013).

In his famous literature on the evolutionary biology of the humanity of Earth, Fazale Rana (2005) briefly elaborated an analysis of Y chromosome variants for Y chromosome Adam, the evolutionary age of whom has been determined between 50,000 to 60,000 YBP. Naturally, this result rendered the evolutionary biologists to construe that in the temporal scale of evolution, Y chromosomal Adam is considered being much younger than mitochondrial eve. In the years of 2011, Fulvio Cruciani and his research colleagues have published their meticulous work on Y chromosome variants of Y Chromosomal Adam entitled "*A revised root for the Human Y Chromosomal Phylogenetic Tree: The origin of patrilineal diversity in Africa*" in the reputed research journal: American Journal of Human Genetics, where they have estimated the evolutionary age or emergence of Y MRCA Adam around 142,000 +/- 16,000 YBP. Based on the analysis of a large chunk of a regular Y chromosome and rare Y chromosome variants, Wei Wei et al (2013) had published a research article, entitled "A calibrated human Y-chromosomal phylogeny based on resequencing," which was published in the science journal, called Genome Research where researchers had claimed the emergence of Y chromosomal Adam has supposedly taken place between 101,000 – 115,000 YBP. So the contemporary works on the Y chromosome variants of Y chromosomal Adam have not only refuted the initial evolutionary age dating process of Y chromosomal Adam, which was published earlier in 2005 but put back in the identical evolutionary age of mitochondrial eve. In 2013, several researchers, who claimed that the emergence of Y-chromosomal Adam supposedly took place approximately between 581,000 YBP and 237, 000 YBP, it's supposed to be 338,000 YBP (Rana, April 4, 2013). So, this evolutionary age of Y-chromosomal Adam, which has found to be 338,000 YBP, indicated that the Y chromosomal Adam has emerged 200,000 years earlier than the anatomically modern human species emerged on the Earth (Rana, April 4, 2013).

The molecular anthropologists engaged in analyzing the DNA sequences of the Y chromosome and mitochondrial chromosome to sort out three main issues to unfurl the evolutionary history of humanity (Carter et al, 2018):

1. To Determine whether the perceptional identity of Adam and Eve is tenable from evolutionary aspects in the history of humanity on Earth or not.
2. Whether the biblical timeframe or the evolutionary timeframe would be a feasible interpretation of depicting the emergence of

Adam or Eve or the temporal history of these primordial, ancestral root-stock of humanity likely to be evolved somewhere between these time-frames.

3. The flexible lengths of Y chromosomal and mitochondrial family trees have projected that the rate of accumulation of mutation is not consistent in the Spatio-temporal scale. So, such observation primarily jeopardizes the credibility of the "molecular clock hypothesis" and its secondary extrapolations like the "Out of Africa" hypothesis, related to the maiden attempt of migration of our ancestors in historical time.

The recent advancement in DNA sequencing have led the molecular anthropologists to the door-step of the treasure trove of genomic data of the entire human chromosomal databases, like 1000 Genomes Project (1000 Genome Project Consortium, 2015), Simons Genome Diversity Project (Simons Genome Diversity Project (Mallick et al, 2016) and the contemporary ongoing Hong Kong Genome Project (Staff Reporter, Genome Web, 16th January 2019). More than 13,000 Y-chromosome SNVS or single nucleotide variations, the diverse range of data in the domain of human genomics, which was recently published by Hallast et al (2015), have unfurled the untold history of humanity on Earth. The molecular anthropologists can pull up several unexplored information of the Y chromosomal data, globally dispersed. The meticulous sequencing investigations of Poznik et al (2016) on 60,000 SNVs, 1400 indels, 110 copy number variations or CNVs and 3,200 STRs or short tandem repeats have exclusively made from 1,200 full-length chromosomes, sequenced by 1000 Genomes Project Consortium (2015) as it found to unfurl the genetic history of humanity on Earth. Though, the population geneticists and molecular anthropologists accepted the limitations in the preliminary level of investigations to identify the haplogroups or haplo- evolutionary-lineages, which is restricted in recognizing a small set of discriminating alleles. The scientists demonstrated that inclusion of related molecular sequences in natural haplogroups leading the reconstruction of the ancestral sequence of each group and the initiation of such earliest efforts, scientifically recognized as "paleogenetics," dates back to 1960s, initiated by Pauling and Zuckerlandl (1963). The initial effort of such investigations has started with an exploration of the "typing method for mitochondria involved restriction endonuclease digestion" (Johnson et al, 1983). Substantial progress has been witnessed in recognizing the haplogroups observed with successful sequencing of HVRI and HVRII hypervariable regions in the course of cladistics analysis (Handt et al, 1998). So, the molecular anthropologists recognized that the documentation of mutational changes in the Y and mitochondrial chromosomes could empirically be recognized as well as interpretation of the genetic history of modern human species, could have been traced back by identifying the distinct form of haplogroups (Carter et al, 2018).

Though the Y chromosome typing started with a sampling of STRs (Short Tandem Repeats) and analyses of Alu insertion events, it gradually progressed with analyses of SNVs (Single Nucleotide Variants) and the full-fledged SNV data eventually available for the Y chromosome and that lead changing of haplogroups identification protocols, that restricted analyses of a set of limited alleles to review of the SNV data of entire chromosome (Romuladi et al, 2002; Purps et al, 2014; Smith, 2015). However, the molecular anthropologists have demonstrated that in their mitochondrial and Y-chromosomal gene trees, each of the branches of a tree historically represents a metaphoric evolutionary predecessor that carry forward one or more "de novo mutations" to its descendants (Carter et al, 2018). Furthermore, the scientists have defined the emergence of each branch of the evolutionary tree as the empirical reference of an individual in terms of its evolutionary age and its identity of its ancestor and descendants. A number of overlapping factors like gene duplication, deletion, genomic realignments, gene conversion, inconsistent rate of mutations, etc. have rendered the phylogenetic reconstruction of trees difficult so that led the reconstructions of historical sequence untenable (Carter et al, 2018). The earlier efforts of such phylogenetic reconstructions in the late nineties was started with the adoption of the parsimony method, introduced by Jermann et al (1995) which has been upgraded to maximum likelihood methods, employed by Pupko et al (2000). Since 2000, the parsimonious tree reconstruction approach has been replaced by Bayesian methods introduced by Huelsenbeck and Bollback (2001). Though the molecular hypothesis has primarily explored by Zuckerkandl and Pauling (1962), whenever they have studied the presence of several heterogeneous amino acids, found in the hemoglobin molecules in different lineages, changes linearly with time and they have drawn a general inference on the monitoring of evolutionary changes of a specific protein, that remained constant for some time, in different evolutionary lineages. Since, the late nineties, the popularization of usage of "molecular clock" has found to be emerging as a prime protocol in the reconstruction of ancestral molecular sequence and interpretation of the evolutionary history of any entity that has transcended and diverged out through distinctive trails of evolution. Yet, the scientific assumption about the temporal accuracy of any molecular clock that depicting the evolutionary history of any evolutionary lineage and its divergence remained to be a hot potato from its very beginning and it has been raised by different molecular biologists over different periods (Tomkins and Bergman, 2015; Jeanson, 2016). However, the contemporary trend of using the molecular clock hypothesis in the reconstruction of a diverse range of phylogenetic trees comprised of diverse evolutionary lineages is found to be still on. The molecular anthropologists have a stereotypic bias in the interpretation of phylogenetic trees, reconstructed with the help of molecular clock models as they have considered the dissimilar sequences as the ancestral types without considering any possibilities that evolutionary age of those sequences might be same with other sequences in its ambiance and the possible reason of such dissimilar appearance might have been triggered by the accelerated rate of mutations the concerned sequences have gone through. However, the molecular phyllogeneticists have acknowledged the molecular and functional complexities for the special chromosomes like Y Chr and Chr M, the unique chromosomal entities with non-recombining DNAs with uniparental inheritance as the reconstruction of ancestral sequences seems to be comparatively simple with the molecular sequences embedded in such chromosomal entities (Carter et al, 2018). The most intriguing part of such reconstruction of ancestral sequences with Y-chromosomal and M-chromosomal entities have been laid in its simplistic inherent, genetical advantages like no obligatory compulsions for identifying the matrix of synteny (the location of a chromosome, in which the alignment of genes found to be sequentially arrayed) in diverse evolutionary lineages. So, it has given enough opportunities to the molecular phylogeneticists to explore genetic histories of the patrilineal and matrilineal lineages of anatomically modern human species in the world.

In their recent investigations, John C. Sanford of Cornell University, USA along with Robert W. Carter and Stephen Lee, two contemporary molecular anthropologists and phylogeneticists have shared their research observations to trace back the genetical history of ancestral lineages of modern human species titled "*An overview of the independent histories of the human Y chromosome and the human mitochondrial chromosome,*" which was published in "*The proceedings of the International Conference on Creationism,*" the published data on the DNA sequence of the Y chromosome, mitochondrial and chromosome 22 sequences have been used from 1000 genome project and Simons genome diversity project (1000 genome project consortium, 2015; Mallick et al, 2016; Carter et al, 2018). So Carter et al (2018) have reviewed the secondary data on Y and mitochondrial sequence assorted it and created a "full distance matrix" for the 1000 genomes Y and mitochondrial sequence data to reconstruct "The neighbor-joining trees" by using the special algorithm called Molecular Evolutionary Genetics Analysis or MEGA, version 7, which has initially developed by Masatoshi Nei, along with research associates Sudhir Kumar and Kochira Tamura (Kumar et al, 1993). Carter et al (2018) had completed the reconstruction of the ancestral sequence for every haplogroup lineages in the format of a simple decision tree, identical to that of Pauling and Zuckerlandl (1963), where the state of the allele in a group has been compared to the state of the alleles, falling outside the group, to secure the justified assignments of the ancestral alleles.

The unique approaches of reconstruction of the simple ancestral state, made by Carter et al (2018) have amazed the contemporary phylogeneticists like Poznik et al (2016) as they have commented that the "*pooling sequences into sub-trees*" should computationally, considered to be an efficient process by the presence of homoplastic traits in it. Carter et al (2018) have claimed that their experimental model of reconstruction of the ancestral state is unique as it has the provision of sorting and review of data in different stages of its progress, leading the option open for cross-checking multiple conclusions. As it has been observed by the molecular phylogeneticists (Hanson-Smith et al, 2010; Groussin et al, 2015) that the distinct phylogenetic protocols (ranging from parsimony, maximum likelihood, and Bayesian approaches) lead yielding slightly different results in the respective phylogenetic reconstructions, so it is convincible that the tentative results yield out of ancestral reconstruction, supposedly little different from the other results (Carter et al, 2018).

However, in their meticulous research experiments, the molecular phylogeneticists, as well as molecular anthropologists have analyzed the molecular sequence of chromosome Y and chromosome M to decipher the genetic and phylogenetic histories, done individually and in parallel and that helped the scientists to contemplate a feasible depiction of the phylogenetic histories of the derived group of modern human species, who have descended out of the cradle of the ancestral human root-stock, once evolved around tens of thousands of years ago and the key observations of those molecular phylogeneticists are as follows (Carter et al, 2018):

1. The empirical trend of the results, achieved from the analysis of Chr. Y and Chr. M, have revealed that both show identical trends, and both datasets indicate an exponential surge of the population with an emergence of several branches that branched out from the root-stock of phylogenetically close related individuals. In their preliminary observation on the analysis of the sequences, of the phylogenetic trees of Y Chr and Chr M, the scientists have noticed that there is a starburst pattern of asymmetrical arrangement of sequences (star-like topology), prevailing in those respective trees and both patterns are found to be identical, as par visual examination. So, scientists have contended that the exponential surge of the population helped them to trace back the mutational changes the derived group has through, descended out of their respective ancestors as most of the mutational changes are instantaneously drifted out of the population and difficult to track down, where the size of the populations has found to be fragile or diminishing in nature (Rupe and Sanford 2013). The critical reviews of the reconstruction of the Y Chr tree helped scientists to trace out the phylogenetic identity of the contemporary haplogroups and it ascertained further the emergence of 11 major haplotypes, supposedly emerged out of this patrilineal evolutionary predecessor (Carter et al, 2018). Whereas the phylogenetic analysis of sequences of M Chr tree has indicated that there are around 16 haplotypes were found to be emerged out of this matrilineal evolutionary mother stock along with a small number of haplogroups reportedly merged in a larger group and those are assumed to be descended out of the identical predecessors (Carter et al, 2018).

2. The critical observation on the dynamics of mutational changes of both phylogenetic trees (of Y Chr and M Chr) has rendered scientists to construe that the rate of mutational changes is not found to be consistent or the same throughout the period of emergence of the derived phylogenetic lineage out of the ancestral stock (Carter et al, 2018). The scientists have argued that the occurrence of mutations happened in parallel in the evolutionary lineages, witnessed among thousands of locations of Y chromosomes and

hundreds of locations in mitochondrial data. Homoplasy has scientifically been recognized as a character, that has been shared by a set of species but, it has found to be absent in their common ancestors. The scientists have figured out the genetic factors triggered the occurrence of homoplasy at the molecular level mainly due to the involvement of an allele, integrated with the main phylogenetic branch point and it has also appeared in a single individual in an unrelated evolutionary group (Carter et al, 2018). Furthermore, the molecular phylogeneticists have argued that either occurrence of homoplasy most likely to be happened due to the sequencing errors (which is most unlikely), or it happened due to several repeated mutations or gene conversions, likely to be happening at the same location in the disconnected lineages. The researchers have further observed that several "ancestral mitochondrial allele calls" have been found to be different from the contemporary phylogenetic work (Behar et al, 2012) but the critical review of 1000 genome datasets revealed that there have no "ancestral allele calls," that disputed the main branches of the Y chromosomal phylogenetic tree (Scozzari et al, 2014).

3. The phylogeneticists have referred the identical starburst pattern displays, found in the Y Chr and M Chr phylogenetic trees, found around the ancestral individuals, belonging to the historical populations, once lived in Middle East Asia (Carter et al, 2018). The molecular phylogeneticists have analyzed the distance matrix for Y chromosome haplogroup founders and that has revealed that there are three large clusters of phylogenetically close-related Y chromosome groups and these are L/T, N/O, and Q/R as well as the ancestors of G, H and I/J and the ancestors of D/E and C, constituting the major lineages of the Y chromosome (Carter et al, 2018). Whereas, the analysis of distance matrix for the mitochondrial haplogroup founders form multiple evolutionary lineages like B, F, H/V/R, J/T, and U/K and all these lineages have been diverged out of from an ancestral lineage, from the macro-haplogroup R (Carter et al, 2018). Based on their critical review, the molecular anthropologists have drawn the inference that the ancestral lineage of clade L3 (i.e. out of Africa) was found to be identical to the major haplogroup M (Campbell and Tishkoff, 2008; Carter et al, 2018). But the critical review of researchers has shaken the foundation of "Out of Africa" hypothesis as researchers have failed to determine the noticeable evolutionary period in between the emergence of the group that assumed to have left Africa and the group that remained in Africa (Campbell and Tishkoff, 2008; Carter

et al, 2018). The molecular phylogeneticists have presented a unique hypothetical scenario about the population size of the ancestral human population and they have contemplated further that if all haplogroups diverged out of the ancestral population of ~10,000 individuals, the founding haplogroup lineages would never be phylogenetically close related.

4. By analyzing the sequence datasets, the molecular phylogeneticists have tracked down the molecular sequences of the archaic and historical individuals that yield each branch in Y Chr and M Chr family trees (Poznik et al, 2013; et al, 2016). The molecular phylogeneticists have critically studied the phylogenies of the asymmetrical topologies of both phylogenetic trees (Y-chromosomal and M chromosomal), that look identical. After studying the typical star-burst topology of both phylogenies, the scientists have construed that tightly-clustered, radiating, starburst pattern of arrangement of haplogroup founders, indicated that the emergence of all these ancestral haplogroups happened in a short period from a small size of the population as a result of exponential population surge (Hey, 2005; Carter et al, 2018). The molecular phylogeneticists have referred the genomic data in support of their preliminary observation. Furthermore, the examination of Y chromosomal data indicated further that most individuals, belonging to the anatomically modern human species, supposedly descended from a small population of closely-related individuals, like-wise the identical analysis, from M chromosomal data ascertained the same observation. Carter et al (2018) shared further that the rate of mutation in Y and M chromosome is found to be less than one per chromosome per generation. So, the back-tracking of mutational changes as well as reverse running of the molecular clock rendered scientists to trace back the post-flood dispersion or surging of the ancestral human population that has assumed to recover the genesis flood induced population bottleneck or drastic shrinkage of ancestral human population and it is noticeable that backtracking does not help to get the same tree, which emerged earlier (Carter et al, 2018). However, the identical asymmetrical sequence in both trees indicated unfurling towards similar consequences of the evolutionary history of human species, evolved in historical time. The molecular phylogeneticists further contended that their study has indicated that a long branch, that distinguishes the Eurasian groups from the evolutionary groups of Africa, was found to be phylogenetically close related to Africans, though the structural and genetical profiles of the individuals belonging to these both

evolutionary groups was found to be distinct (Poznik et al, 2013; Carter et al, 2018).

5. The molecular phylogeneticists figured out that within a feasible range of the margin of error, they figured out an approximation of the sequence of Y chromosome Adam and or Noah and Mitochondrial Eve (Carter et al, 2018). The main impediment of this genomic investigation is integrated into the poor detection result of copy number variations and the scientists failed to establish the comparison of a high degree of divergence data to the low level of sequence coverage (Poznik et al, 2013; Carter et al, 2018).

6. Robert W Carter, Stephen S. Lee, and John C. Sanford have considered a reasonable contemplation, helped researchers to estimate the evolutionary time of Mitochondrial Eve, and Y chromosome and Noah (Carter et al, 2018). From an evolutionary time-frame, the molecular phylogenetists contended that both Adam and Noah, both of the biblical characters, have lived less than 10,000 years ago (Carter et al, 2018). The molecular phylogeneticists analyzed the Y chromosomal sequences of 25 individuals of 69 high coverage genome dataset and 55 genome subset and built a phylogenetic tree and compared the genomic data between the Complete Genomics and scientists have detected around 5,000 SNVs (Drmanac et al, 2010; 1000 Genomics Consortium, 2015; Poznik et al , 2016). Furthermore, researchers have identified 39 points among 23 genomic locations where the data sets of 1000 Genomics and Complete Genomics have contradicted one another and half of the 23 genomic locations i.e. 12/ 23 or 0.44, recognized as private alleles, are found to be missing per person (1000 Genomics Consortium, 2015). The researchers also noticed that 15 out of these 23 genomic locations in the larger data set are found to be variable within one of the major haplogroups, whereas 2 of these 23 are found to be fixed in major haplogroups and rest of the 6 are found to be homoplastic in traits (Carter et al, 2018).

7. In course of overviewing the independent histories of the human Y chromosome and the human mitochondrial chromosome, the molecular phylogeneticists have acknowledged that accumulation of recurrent mutations is a common evolutionary trend in evolutionary genetics and it normally happened in the epigenetic CpG sites[5]

---

[5] CpG Site: CpG or CG site is the specific region of DNA, where Cytosine nucleotide has been linearly sequenced with Guanine nucleotide to the 5' to 3.' The

(Carter et al, 2018). Furthermore, scientists have observed that accumulation of mutation does not happen independently without any influence of environmental influence, rather an accumulation of mutation happened in distinct evolutionary lineages in parallel. The molecular phylogeneticists acknowledged that the time gap between the ancestor and its derived haplogroup has remained to be constant for individuals of that group. The scientists have further observed that in each founder haplogroup there were derived individuals with molecular sequences, phylogenetically belonging outskirts of its living descendants. After a critical review of the sequence data of the Y Chr and M Chr phylogenetic tree, scientists have contemplated that mutational divergence[6] is less likely to be constant in all evolutionary lineages. The evolutionists have believed with some exception of individuals, belonging to haplogroup A, haplogroup B has been considered as the ancestor of every man. Surprisingly, scientists have noticed that B individuals had ranked much higher than expected. So, that made the molecular phylogeneticists to construe that the constancy of the rate of mutational divergence has not been maintained among the evolutionary lineages, less likely. Theoretically, the all evolutionary pairs of Y Chr and Chr M, supposedly, "equally diverged" from their common ancestor but in reality, such evolutionary assumptions have not been reflected among many family pairs as they have emerged with different degrees of divergence. In support of this observation, the molecular phylogeneticists have given the example of haplogroup pairs A1 and B of Y chromosome, found to be significantly less diverged from their ancestral lineage A1, than the members of the other haplogroups, whereas the haplogroup C has found to be more diverged than its ancestral lineage C in respect to the other descendant haplogroups (Carter et al, 2018). Nonetheless, whatever the ancestral lineage would have been chosen for comparison the differential divergence rates were found to be prevailing among the ancestral lineage and its descendants, so in

---

Cytosine of the CpG site has been methylated to form 5 methylcytosine and the methylation of cytosine in a gene changes the expression of gene, and such methlation-induced gene regulation deals with gene regulation action I the arena of epigenetics.

[6] Mutational Divergence: By means of mutational divergence, more than one populations of any ancestral species have gone through the accumulation of sudden genetic changes or mutations through time and that rendered the populations reproductively isolated and emergence of new sub-populations, those are found to be suitable in the ecologically peripheral environment.

terms of molecular clock analogy, the molecular clock does not tick consistently in the family lines of every individual.

8. Molecular phylogeneticists demonstrated that the newly derived evolutionary lineages diverged out of the ancestral lineage dichotomously. When more than two evolutionary lineages emerged out of a common ancestral rootstock, it's called polytomy (Lafond et al, 2014; Carter et al, 2018). In course of drifting away of newly acquired mutations drifted out of the gene pool, the chance of emergence of an any derived evolutionary lineage out of the ancestral one is considered to be a rare occurrence, so the possibility of identifying a founder evolutionary lineage, that would have rendering divergence of more than two derived lineages with a substantial accumulation of mutational divergence, is considered to be less likely (Fu et al, 2013; Carter et al, 2018). But polytomic phylogenetic evidence is not unusual though as the analysis of a phylogenetic tree of Y chromosomal sequence, shows that it has gone through the 3 way polytomic orientation where the ancestral macro-haplogroup has given rise polytomic divergence of K to three derived lineages: L/T, O/N, and Q/R haplogroups (Hallast et al, 2015). The polytomic displays are there in mitochondrial data displays, then the analysis of the dataset of the Y chromosome data. Polytomic divergence of derived evolutionary lineages out of the ancestral lineage is visually indistinguishable unless critical observation has been made to distinguish that the tightly-adhered derived branches are found to be arrayed close to each other, even less than one mutation unit and visibly they are indistinguishable (Fu et al, 2013; Carter et al, 2018). Hence, scientists contemplated that the branching off of more than one female lines out of one or more than one founding females in course of evolutionary divergence of modern human species.

9. So, the next concern intrigued molecular phylogeneticists to figure out the causes, responsible for influencing the mode of variable mutation accumulation. The long term evolution experimental project on its genetic divergence, which has got an acronym of LTEE bacterial experiment with *Escherichia coli* has been successfully made by Tenaillon et al (2016), where a single strain of *E.coli* has separated into 12 evolutionary lineages over time. Furthermore, the molecular geneticists have successfully cloned two clones of each of these 12 lineages, after completion of 50,000 cycles and researchers noticed that six out of 12 lineages have accumulated 43.1 mutations over 50,000 generations but the remaining six lineages have an exponential surge of mutations at the

rate of 1100 to 2500 accumulated mutations (Tenallion et al, 2016). So, it has been observed that LTEE bacterial experiment shows that in a given time, two different lineages, that diverged out of a common lineage have maintained two differential magnitudes of mutation, where the rate of accumulation of mutation in hyper-mutative strain is found to be considerably higher than a normal one (Tenaillon et al, 2016). The molecular phylogeneticists have strongly advocated the LTEE evolutionary model in defining the variable rate of mutational accumulation in the human genome about the divergence of new haplogroups out of the ancestral cradles in human species. The molecular anthropologists analyzed the pattern of variable mutation accumulation in the rare African lineages in comparison to all other evolutionary lineages but the statistical interpretation of such mutation accumulation patterns in driving the evolutionary divergence of derived lineages was found to be inconclusive (Carter et al, 2018). In an experiment, Rozen et al (2003) contended that around 600 nucleotides of each newborn male went through gene conversion in the Y chromosome and the conversion rate primarily varies by factors like sex and age of the individual and it has secondarily influenced by heterozygosity (Trombetta et al, 2014). As scientists have noticed that heterozygosity influenced the chromosomal recombination and it rendered the molecular geneticists to construe further that gene conversion is a tenable explanation to define the tentative accumulation for a variable mutation in human (Duret and Galtier, 2009; Trombetta and Cruciani, 2017). The molecular anthropologist tendered the example of the extent of heterozygosity in terms of the gene conversion of the ancient ethnic group of human species as the scientists have observed that in the chr.no. 22 of 504 individuals of the African population were found to be heterozygous nature at 3.78% of all variable alleles whereas in the 502 individuals of European populations the rate of heterozygosity was found to be at 2.91 % level (Carter et al, 2018; Korunes and Noor, 2019). The rate of chromosomal recombination of African populations was found to be noticeably higher than non-Africans (Hinch et al, 2011). The example of such possible cause for variable mutation accumulation rates in human has found to be intriguing but it is too early to claim that gene conversion is solely related to the extent of heterozygosity in modern human species unless gene conversion is integrated to recombination rates that could define the mutation dynamics in terms of divergence of evolutionary lineages in modern human species (Carter et al, 2018; Korunes and Noor, 2019).

10. So the final challenge of this experimental demonstration of narrating the independent histories of the Y chromosome and M chromosome of humanities was the estimation of the age of the primary root sequences for those respective chromosomes, which primarily requires modeling of a molecular clock that can decipher the rate of mutation accumulation among the different evolutionary lineages and to do an age estimation, it requires a consideration of a big margin of error. So the investigation of this evolutionary investigation comes straight to finding out the tentative rate of mutation of the Y chromosome. Xue et al (2009) have figured out a rate of 3.0x10-8 per site, per person, per generation, where the assumption has been made based on the evolutionary divergence of humans, split out with chimpanzee, shared the common ancestral rootstock around 6.5 mya (Gurven and Davison, 2019). Very recently, Skov et al (2017) experimentally determined the rate of mutation 3.14 x 10-8 per site, per person, per generation for the X-degenerate section of the Y chromosome (where the generation time has been calculated of 36 years). So, let's go to the next level to figure out the mutation rate of the M chromosome. The molecular geneticists have observed that the rate of mitochondrial mutational variations prevailed and depends on several issues: area sampled, types of mutations, etc. The lack of standard measure has appeared to be a problem in the molecular clock to monitor mitochondrial studies (Loogvali et al, 2009). Henceforth, Soares et al (2009) attempted to segregate the mitochondrial genome into eight sections to figure out the tentative rate of mutation in it and which was found to be 1.7 x 10^{-8} mutations in each site, per year (at the rate of one mutation in every 3,624 years, which is assumed to be too low. This estimation was made under the influence of phylogenetic contemplation that projected the split of humans and chimpanzee around 6.5 mya (Gurven and Davison 2019; Carter et al, 2018). So, such phylogenetic mutational rate has been discarded by the contemporary molecular phylogeneticists, rather a mutational estimation, done by the genealogical method that has been introduced by Madrigal et al (2012) has been appreciated, as it has projected the mutational rate of 1.24 x 10-6 per site, per year in the specific location, recognized as HVSH or second hypervariable section of M Chr. This mutational rate, which is approximately 1 mutation every-other generation, is found to be tenable than the projection of mutation rate of M Chr, made by Soares et al (2009). Carter et al (2018) have determined that genome-wide there were 25.3 times more mutations than in HVSII and it has also projected genome-wide 8.5 times more mutations than in the control region.

> Though interpolation of such data helped further to estimate the evolutionary age of the haplogroup ancestors, with a degree of uncertainty.

On the way of approximation of the primary root sequences for Y Chr and M Chr, evolutionists have enjoyed a strategic advantage in using the chimpanzee as a baseline or rooting lineage as it has been assumed that the ancestor of modern human supposedly possesses chimpanzee-like sequences as Chimp-genome has assumed to carry ancestral allele (Park, 2015; Carter et al, 2018). It does not indicate that a part of the human population closely resembled chimpanzees than others but it indicates that a few evolutionary lineages branched off a long time ago. On the other hand, Carter et al (2018) tried to narrate I from a biblical perspective and readily accepted that its difficult to claim undoubtedly that the journey of humanity started with a single Y Chr lineage or if the sons of Noah are identical or looking different. Furthermore, the scientists observed that the same observation is applicable for M Chr too (Carter et al, 2018). As the biblical identity of the three daughters-in-law of Noah is not very clear whether they are closely related or not, so it is difficult to predict the root of the tree as well as predicting about the primary root sequences for Y Chr or M Chr (Ark Encounter, 12th July, 2018; https://arkencounter.com/blog/2018/07/12/meet-ladies-noahs-family/) .

Carter et al (2018) have contended in their phylogenetic investigations, they have observed "…in both Y chromosomes and mitochondrial DNA supports a single primary dispersion of humanity in the recent past. Some call this dispersion 'Out of Africa' the independent history of humanity Y Chr and Mit DNA analysis indicates towards a single primary dispersion of humanity in the recent past. Some call this dispersion 'Out of Africa' We call it 'Out of Babel.'" The scientists have observed that in both trees of Y Chr and M Chr, the major number of evolutionary lineages have been traced to similar to the ancestral individuals so those could be brothers/sisters or cousins in relation. Carter et al (2018) observed "If the mutation rate for Y Chr and M Chris less than one per generation, it might not be possible to capture all linage-forming events, but the fact remains that the individuals who gave rise to the major clades were not far apart in time….The chances are vanishingly small that in a large population any two closely-related people would go on to have millions of ancestors." Yet this is exactly what we see happening, many times, in both of the Y Chr and M Chr trees.

Furthermore, Carter et al (2018) contended that exponential growth of the human population assumed to have happened at the beginning of the history of humanity, perhaps initiated with a few families or small tribes. It has also been contended by researchers that such observation has hard to find to fit in the evolutionary model, rather it has found to fall out of the biblical model with rapid growth from a single and small population (Carter and Hardy, 2015; Carter and Powell, 2016).

The molecular phyllogeneticists and molecular anthropologists tried to narrate the history of humanity by making an effort to trace back the major evolutionary lineages as the substantial evidence proved that humanity had supposedly gone through the population bottleneck, subsequently followed by a surge of the population (Keinan and Clark, 2012). Hence, it has found to be fit in the Genesis Flood/Babel model under the theological perception of Creationism, rather than the principles of evolution.

References

Behar, D.M., van Oven, M., Rosset, S., Metspalu, M., Loogväli, E.V., Silva, N.M., Kivisild, T., Torroni, A., and Villems, R. (2012). A Copernican reassessment of the human mitochondrial DNA tree from its root. *American Journal of Human Genetics* 90:675-684.

Campbell, M. C., and Tishkoff, S. A. (2008). African genetic diversity: implications for human demographic history, modern human origins, and complex disease mapping. *Annu Rev Genomics Hum Genet.* 9:403-33. doi: 10.1146/annurev.genom.9.081307.164258.

Cann, R.L., Stoneking, M. and Wilson, A.C. (1987). Mitochondrial DNA and human evolution. Nature. 325: 31-36.

Carter, R. and Hardy, C. (2015). Modelling biblical human population growth. *Journal of Creation* 29(1): 72-79.

Carter, R.W., Lee, S.S. and Sanford, J.C. (2018). An overview of the independent histories of the human Y-chromosome and the human mitochondrial chromosome. In *Proceedings of the Eighth International Conference on Creationism*, ed. J.H. Whitmore, pp. 133–151. Pittsburgh, Pennsylvania: Creation Science Fellowship.

Carter R., and Powell, M. (2016). The genetic effects of the population bottleneck associated with the Genesis Flood. *Journal of Creation* 30(2):102–111.

Coogan, M.D. (2001). *The Oxford History of the Biblical World.* Oxford University Press, p-508.

Duret, L., and Galtier, N. (2009). Biased gene conversion and the evolution of mammalian genomic landscapes. *Annu Rev Genomics Hum Genet.* 10:285-311. doi: 10.1146/annurev-genom-082908-150001.

Drmanac, R., Sparks, A.B., Callow, M.J., Halpern, A.L., Burns, N.L., Kermani, B.G., Carnevali, P., Nazarenko, I., Nilsen, G.B., Yeung, G., Dahl, F., Fernandez, A., Staker, B., Pant, K.P., Baccash, J., Borcherding, A.P., Brownley, A., Cedeno, R., Chen, L., Chernikoff, D., Cheung, A., Chirita, R., Curson, B., Ebert, J.C., Hacker, C.R., Hartlage, R., Hauser, B., Huang, S., Jiang, Y., Karpinchyk, V., Koenig, M., Kong, C., Landers, T., Le, C., Liu, J., McBride, C.E., Morenzoni, M., Morey, R.E., Mutch, K., Perazich, H., Perry, K., Peters, B.A., Peterson, J., Pethiyagoda, C.L., Pothuraju, K., Richter, C., Rosenbaum, A.M., Roy, S., Shafto, J., Sharanhovich, U., Shannon, K.W., Sheppy, C.G., Sun, M., Thakuria, J.V., Tran, A., Vu, D., Zaranek, A.W., Wu, X., Drmanac, S., Oliphant, A.R., Banyai, W.C., Martin, B., Ballinger, D.G., Church, G.M., and Reid, C.A. (2010). Human genome sequencing using unchained base reads on self-assembling DNA nanoarrays. *Science.* 327(5961): 78-81. doi: 10.1126/science.1181498.

Endicott, P., Ho, S.Y., Metspalu, M., and Stringer, C. (2009). Evaluating the mitochondrial timescale of human evolution. *Trends Ecol Evol.*24(9):515-21. doi: 10.1016/j.tree.2009.04.006.

Fu, Q., Mittnik, A., Johnson, P.L.F., Bos, K., Lari, M., Bollongino, R., Sun, C., Giemsch, L., Schmitz, R., Burger, J., Ronchitelli, A.M., Martini, F., Cremonesi, R.G., Svoboda, J., Bauer, P., Caramelli. D,, Castellano, S., Reich, D., Pääbo, S., and Krause, J. (2013). A revised timescale for human evolution based on ancient mitochondrial genomes. *Curr Biol.* 23(7): 553-559. doi: 10.1016/j.cub.2013.02.044.

Groussin, M., Hobbs, J.K., Szöllösi, G.J., Gribaldo, S., Arcus, V.L., and Gouy, M. (2015). Toward more accurate ancestral protein genotype-phenotype reconstructions with the use of species tree-aware gene trees. *Molecular Biology and Evolution.* 32: 13-22. DOI: 10.1093/Molbev/Msu305

Gurven, M., and Davison, R. (2019). Periodic catastrophes over human evolutionary history are necessary to explain the forager population paradox. *PNAS* 201902406. +doi:10.1073/pnas.1902406116

Hallast, P., Batini, C., Zadik, D., Maisano Delser, P., Wetton, J.H., Arroyo-Pardo, E., Cavalleri, G.L., de Knijff, P., Destro Bisol, G., Dupuy, B.M., Eriksen, H.A., Jorde, L.B., King, T.E., Larmuseau, M.H., López de Munain, A., López-Parra, A.M., Loutradis, A., Milasin, J., Novelletto, A., Pamjav, H., Sajantila, A., Schempp, W., Sears, M., Tolun, A., Tyler-Smith, C., Van Geystelen, A., Watkins, S., Winney, B., and Jobling, M.A. (2015). The Y-chromosome tree bursts into leaf: 13,000 high-confidence SNPs covering the majority of known clades. *Mol Biol Evol.* 32(3):661-73. doi: 10.1093/molbev/msu327.

Handt, O., Meyer, S., and von Haeseler, A. (1998). Compilation of human mtDNA control region sequences. *Nucleic Acids Res.* 26(1): 126-9. doi: 10.1093/nar/26.1.126.

Hanson-Smith, V., Kolaczkowski, B., and Thornton, J.W. (2010). Robustness of Ancestral Sequence Reconstruction to Phylogenetic Uncertainty, *Molecular Biology and Evolution*, 27(9): 1988– 1999,

https://doi.org/10.1093 / molbev/msq081

Hardy, C., and Carter, R. (2014). "The Biblical Minimum and Maximum Age of the Earth." *Journal of Creation* 28(2): 89–96.

Hey, J. (2005). On the number of New World founders: A population genetic portrait of the peopling of the Americas. *PLoS Biol* 3:e193.

Hinch A. G., Tandon A., Patterson N., Song Y., Rohland N., et al., (2011). The landscape of recombination in African Americans. *Nature* 476: 170–175.

Huelsenbeck, J.P., and Bollback, J.P. (2001). Empirical and hierarchical bayesian estimation of ancestral states. *Systematic Biology* 50(3):351–366.

Jeanson, N.T. (2016). On the Origin of Human Mitochondrial DNA Differences, New Generation Time Data Both Suggest a Unified Young-Earth Creation Model and Challenge the Evolutionary Out-of-Africa Model.

https://www.semanticscholar.org/paper/On-the-Origin-of-Human-Mitochondrial-DNA-New-Time-a-Jeanson/3857a6e05f2589065c7d3a97bd7244028befea30

Jeanson, N. T. (2019). "Testing the Predictions of the Young-Earth Y Chromosome Molecular Clock: Population Growth Curves Confirm the Recent Origin of Human Y Chromosome Differences." *Answers Research Journal* 12: 405–423.

Jermann, T.M., Opitz, T.G., Stackhouse, J. and Benner, S.A. (1995). Reconstructing the evolutionary history of the artiodactyl ribonuclease superfamily. *Nature* 374:57–59.

Johnson, M.J., Wallace, D.C., Ferris, S.D., Rattazzi, M.C., and Cavalli-Sforza, L.L. (1983). Radiation of human mitochondria DNA types analyzed by restriction endonuclease cleavage patterns. *J Mol Evol.* 19 (3-4): 255-71. doi: 10.1007/BF02099973.

Karafet, T.,M,, Mendez, F. L., Meilerman, M. B., Underhill, P. A. , Zegura, S. L., and Hammer, M. F. (2008). New binary polymorphisms reshape and increase resolution of the human Y chromosomal haplogroup tree. Genome Res. 18(5):830-8. doi: 10.1101/gr.7172008.

Karmin, M., Saag, L., Vicente, M., Wilson Sayres, M. A., Järve, M., Talas, U. G., Rootsi, S., Ilumäe, A. M., Mägi, R., Mitt, M., Pagani, L., Puurand, T., Faltyskova, Z., Clemente, F., Cardona, A., Metspalu, E., Sahakyan, H., Yunusbayev, B., Hudjashov, G., DeGiorgio, M., Loogväli, E. L., Eichstaedt, C., Eelmets, M., Chaubey, G., Tambets, K., Litvinov, S., Mormina, M., Xue, Y., Ayub, Q., Zoraqi, G., Korneliussen, T. S., Akhatova, F., Lachance, J., Tishkoff, S., Momynaliev, K., Ricaut, F. X., Kusuma, P., Razafindrazaka, H., Pierron, D., Cox, M. P., Sultana, G. N., Willerslev, R., Muller, C., Westaway, M., Lambert, D., Skaro, V., Kovačevic, L., Turdikulova, S., Dalimova, D., Khusainova, R., Trofimova, N., Akhmetova, V., Khidiyatova, I., Lichman, D. V., Isakova, J., Pocheshkhova, E., Sabitov, Z., Barashkov, N. A., Nymadawa, P., Mihailov, E., Seng, J. W., Evseeva, I., Migliano, A. B., Abdullah, S., Andriadze, G., Primorac, D., Atramentova, L., Utevska, O., Yepiskoposyan, L., Marjanovic, D., Kushniarevich, A., Behar, D.M., Gilissen, C., Vissers, L., Veltman, J.A., Balanovska, E., Derenko, M., Malyarchuk, B., Metspalu, A., Fedorova, S., Eriksson, A., Manica, A., Mendez, F. L., Karafet, T. M., Veeramah, K. R., Bradman, N., Hammer, M. F., Osipova, L. P., Balanovsky, O., Khusnutdinova, E. K., Johnsen, K., Remm, M, Thomas, M. G., Tyler-Smith, C., Underhill, P. A., Willerslev, E., Nielsen, R., Metspalu, M., Villems, R., and Kivisild, T. (2015). A recent bottleneck of Y chromosome diversity coincides with a global change in culture. *Genome Res.* 25(4):459-66. doi: 10.1101/gr.186684.114.

Keinan, A., and Clark, A. G. (2012). Recent explosive human population growth has resulted in an excess of rare genetics variants. *Science*, 336: 740–743. https://doi.org/10.1126/science.1217283

Korunes, K. L., and Noor, M. A. F. (2019). Pervasive gene conversion in chromosomal inversion heterozygotes. *Mol Ecol.* 8(6): 1302-1315. doi: 10.1111/mec.14921.

Kumar, S., Tamura, K., and Nei, M. (1994). MEGA: molecular evolutionary genetics analysis software for microcomputers. *Comput Appl Biosci.* 10:189–191.

Lafond, M., Chauve, C., Dondi, R., El-Mabrouk, N. (2014). Polytomy refinement for the correction of dubious duplications in gene trees, *Bioinformatics*, 30(17): i519–i526,
https://doi.org/10.1093/bioinformatics/btu463

Loogväli, E-L., Kivisild, T., Margus, T., and Villems, R. (2009) Explaining the Imperfection of the Molecular Clock of Hominid Mitochondria. *PLoS ONE* 4(12): e8260. https://doi.org/10.1371/journal.pone.0008260

Madrigal, L., (Posthumously), L.C., Melendez-Obando, M., Villegas-Palma, R., Barrantes, R., Raventos, H., Pereira, R., Luiselli, D., Pettener, D. and Barbujani, G. (2012), High mitochondrial mutation rates estimated from deep-rooting costa rican pedigrees. *Am. J. Phys. Anthropol.*, 148: 327-333. https://doi.org/10.1002/ajpa.22052

Mallick, S., Li, H., Lipson, M., Mathieson, I., Gymrek, M., Racimo, F., Zhao, M., Chennagiri, N., Nordenfelt, S., Tandon, A., Skoglund, P., Lazaridis, I., Sankararaman, S., Fu, Q., Rohland, N., Renaud, G., Erlich, Y., Willems, T., Gallo, C., Spence, J.P., Song, Y.S., Poletti, G., Balloux, F., van Driem, G., de Knijff, P., Romero, I.G., Jha, A.R., Behar, D.M., Bravi, C.M., Capelli, C., Hervig, T., Moreno-Estrada, A., Posukh, O.L., Balanovska, E., Balanovsky, O., Karachanak-Yankova, S., Sahakyan, H., Toncheva, D., Yepiskoposyan, L., Tyler-Smith, C., Xue, Y., Abdullah, M.S., Ruiz-Linares, A., Beall, C.M., Di Rienzo, A., Jeong, C., Starikovskaya, E.B., Metspalu, E., Parik, J., Villems, R., Henn, B.M., Hodoglugil, U., Mahley, R., Sajantila, A., Stamatoyannopoulos, G., Wee, J.T., Khusainova, R., Khusnutdinova, E., Litvinov, S., Ayodo, G., Comas, D., Hammer, M.F., Kivisild, T., Klitz, W., Winkler, C.A., Labuda, D., Bamshad, M., Jorde, L.B., Tishkoff, S.A., Watkins, W.S., Metspalu, M., Dryomov, S., Sukernik, R., Singh, L., Thangaraj, K., Pääbo, S., Kelso, J., Patterson ,N., and Reich, D. (2016). The Simons Genome Diversity Project: 300 genomes from 142 diverse populations. *Nature.* 538 (7624): 201-206. doi: 10.1038/nature18964.

Park, L. (2015). Ancestral alleles in the human genome based on population sequencing data. *PLoS One.* 10(5):e0128186. doi: 10.1371/journal.pone. 0128186.

Pauling, L., and Zuckerkandl, E. (1963). Chemical paleogenetics: molecular "restoration studies" of extinct forms of life. *Acta Chem Scand [B]* 17:9–16.

Pupko, T., Shamir, I.P.R. and Graur, G. (2000). A fast algorithm for joint reconstruction of ancestral amino acid sequences. *Molecular Biology and Evolution* 17(6):890–896.

Purps, J., Siegert, S., Willuweit, S., Nagy, M., Alves, C., Salazar, R. et al. (2014). A global analysis of Y-chromosomal haplotype diversity for 23STR loci. *FSI:Genetics* 12:12–23.

Poznik, G.D., Henn, B.M., Yee, M.C., Sliwerska, E., Euskirchen, G.M., Lin, A.A., Snyder, M., Quintana-Murci, L., Kidd, J.M., Underhill, P.A., and Bustamante, C.D. (2013). Sequencing Y chromosomes resolves discrepancy in time to common ancestor of males versus females. *Science.* 341(6145): 562-5. doi: 10.1126/science.1237619.

Poznik, G.D., Xue, Y., Mendez, F.L., Willems, T.F., Massaia, A., Wilson Sayres, M.A., Ayub, Q., McCarthy, S.A., Narechania, A., Kashin, S., Chen, Y., Banerjee, R., Rodriguez-Flores, J.L., Cerezo, M., Shao, H., Gymrek, M., Malhotra, A., Louzada, S., Desalle, R., Ritchie, G.R., Cerveira, E., Fitzgerald, T.W., Garrison, E., Marcketta, A., Mittelman, D., Romanovitch, M., Zhang, C., Zheng-Bradley, X., Abecasis, G.R., McCarroll, S.A., Flicek, P., Underhill, P.A., Coin, L., Zerbino, D.R., Yang, F., Lee, C., Clarke, L., Auton, A., Erlich, Y., Handsaker, R.E.; 1000 Genomes Project Consortium, Bustamante, C.D., and Tyler-Smith, C. (2016). Punctuated bursts in human male demography inferred from 1,244 worldwide Y-chromosome sequences. *Nat Genet.* 48 (6): 593-9. doi: 10.1038/ng.3559.

Rana, F. (April 4, 2013). When did Mitochondrial Eve and Y Chromosomal Adam Live?

https://reasons.org/explore/publications/articles/when-did- mitochondrial-eve-and-y-chromosomal-adam-live

Rana, F. and Ross, H. (2005). Who was Adam? : A creation model approach to the origin of Man. Navpress publishing House. p.299.

Romualdi, C., Balding, D., Nasidze, I.S., Risch, G., Robichaux, M., Sherry, S.T., Stoneking, M., Batzer, M.A., and Barbujani, G. (2002). Patterns of human diversity, within and among continents, inferred from biallelic DNA polymorphisms. Genome Res 12:602–612.

Rozen, S., Skaletsky, H., Marszalek, J.D., Minx, P.J., Cordum, H.S., Waterston, R.H., Wilson, R.K., and Page, D.C. (2003). Abundant gene conversion between arms of palindromes in human and ape Y chromosomes. *Nature.* 423(6942): 873-6. doi: 10.1038/nature01723.

Rupe, C.L. and Sanford, J.C. (2013). Using Numerical Simulation to Better Understand Fixation Rates, and Establishment of a New Principle: Haldane's Ratchet. *The Proceedings of the International Conference of Creationism.* Vol.7, article 32. .

Scally, A., and Durbin, R. (2012). Revising the human mutation rate: implications for understanding human evolution. *Nat Rev Genet.* 13(10):745-53. doi: 10.1038/nrg3295.

Scozzari, R., Massaia, A., D'Atanasio, E., Myres, N. M., Perego, U. A., Trombetta, B., and Cruciani, F. (2012). Molecular dissection of the basal clades in the human Y chromosome phylogenetic tree. *PLoS One.* 7(11):e49170. doi: 10.1371/journal.pone.0049170.

Skov, L., The Danish Pan Genome Consortium, Schierup, M. H. (2017). Analysis of 62 hybrid assembled human Y chromosomes exposes rapid structural changes and high rates of gene conversion. *PLoS Genet* 13(8): e1006834. https://doi.org/10.1371/journal.pgen.1006834

Smith, D.R. (2015). The past, present and future of mitochondrial genomics: have we sequenced enough mtDNAs? *Briefings in Functional Genomics* doi:10.1093/bfgp/elv027.

Soares, P., Ermini, L., Thomson, N., Mormina, M., Rito, T., Röhl, A., Salas, A., Oppenheimer, S., Macaulay, V., and Richards, M. B. (2009). Correcting for purifying selection: an improved human mitochondrial molecular clock. Am J Hum Genet. 84(6):740-59. doi: 10.1016/j.ajhg.2009.05.001.

Staff Reporter, Genome Web, (January 16, 2019). Hong Kong Genome Project to Sequence, Analyze 20K Patients.

https://www.genomeweb. com/ sequencing/hong-kong-genome-project-sequence-analyze-20k-patients # .YNpwJdUzbIU

Tenaillon, O., Barrick, J.E., Ribeck, N., Deatherage, D. E., Blanchard, J.L., Dasgupta, A., Wu, G. C., Wielgoss, S., Cruveiller, S., Médigue, C., Schneider, D., and Lenski, R. E. (2016). Tempo and mode of genome evolution in a 50,000-generation experiment. *Nature.* 536(7615):165-70. doi: 10.1038/nature18959.

The 1000 Genomes Project Consortium. (2015) A global reference for human genetic variation. *Nature* 526: 68–74. https://doi.org/10.1038/nature15393

Tomkins, J. P. and Bergman, J. (2015). Evolutionary molecular genetic clocks—a perpetual exercise in futility and failure. *Journal of Creation.* 29 (2): 26-35.

Trombetta, B., and Cruciani, F. (2017). Y chromosome palindromes and gene conversion. *Hum Genet.* 136(5): 605-619. doi: 10.1007/s00439-017-1777- 8.

Trombetta, B., Sellitto, D., Scozzari, R., and Cruciani. F. (2014). Inter- and intraspecies phylogenetic analyses reveal extensive X-Y gene conversion in the evolution of gametologous sequences of human sex chromosomes. *Mol Biol Evol.* 31(8):2108-23. doi: 10.1093/molbev/msu155.

Wei, W., Ayub, Q., Chen, Y., McCarthy, S., Hou, Y., Carbone, I., Xue, Y., Tyler- Smith, C. (2013). A calibrated human Y-chromosomal phylogeny based on resequencing. *Genome Res.* 23(2):388-95. doi: 10.1101/gr.143198 .112.

Xue, Y., Wang, Q., Long, Q., Ng, B.L., Swerdlow, H., Burton, J., Skuce, C., Taylor, R., Abdellah, Z., Zhao, Y., Asan, MacArthur, D.G., Quail, M. A., Carter, N.P., Yang, H., and Tyler-Smith, C. (2009). Human Y chromosome base-substitution mutation rate measured by direct sequencing in a deep-rooting pedigree. *Curr Biol.* 19(17):1453-7. doi: 10.1016/j.cub.2009.07.032.

Zuckerkandl, E. and Pauling, L.B. (1962). "Molecular disease, evolution, and genetic heterogeneity," in *Horizons in Biochemistry*, eds. M. Kasha & B. Pullman; New York: Academic Press, 189-225.

Chapter 7

In Quest of Links of One Hundred Thousand Genetic "Adams and Eves" and Their Modern Human Descendants Who Lived Over 135,000 Years Ago

The DNA held inside mitochondria can therefore reveal the maternal lineage to ancient Eve, who is believed to have lived in Africa between 99,000 and 148,000 years ago- almost the same time period during which the Y chromosome 'Adam' lived. However, while these genetic studies reveal a common ancestor for all men and a common ancestor for all women, genetic 'Adam and Eve' are not the same as the biblical Adam and Eve as they were not literary the first modern on the planet, but instead just the two out of thousands people alive at the time with unbroken male or female lineages that continue on today. Furthermore, while they may have overlapped in time, it is unlikely that ancient 'Adam' and the ancient 'Eve' lived near each other, let alone and reproduced.

---- *April Holloway*

It is quite thrilling to be conceived with such a perception that human genome, has not only found to be the precious holy grail for molecular phylogeneticists, genome biologists, molecular biologists, paleo-anthropologists and evolutionary biologists exploring the genealogical treasure-troves but it has emerged to be the unique archive for the historian who would like to define and redefine the emerging history of humanity. The extent of that has been extended from the contemporary history of diversifications of the modern human species to the beginning of the evolutionary journey of archaic human lineages, either descended from the evolutionary cradle of great apes or emerged out of the biblical narrative of "Adam and Eve," considered to be the primordial couple on Earth. In 1999, a unique historical narrative of Britain has been presented in "*The Isles: A*

History" authored by Norman Davies, in the backdrop of the portrayal of a magnificent evolutionary journey of humanity through time. The main focus of attraction of this book is confined to a prehistoric human skeleton excavated from Cheddar village, (around 14km north-west of Welles, in the U.K.) which has found to be 8,980 years old and the scientists have successfully examined the mitochondrial DNA sample, extracted from it. So, the primary findings of the researcher regarding the early record of a hominin from such location are stunning. But its turned around to be more exciting when the scientists have engaged in random investigations of DNA sequencing of the small population of this small village and they surprisingly amazed with a DNA match result of Adrian Targett, a school teacher of Cheddar Village School and it mesmerized the community of science that after traversing a long period down the evolutionary trail of 9 millenniums, the vertical inheritance of genes from ancestor to the present day successor has remain been uninterrupted, though the journey of humanity has contemplated passing through a series of genetic bottlenecks in the meantime (Amos and Hoffman, 2010; McKie, February 10, 2018). Though the history of DNA sequencing, investigations was initiated in the late nineties intending to identify the individuals, engaged in some activities, related to Forensic science and determining the confirmatory paternity tests in terms of "DNA fingerprinting," and it started shifting towards the different frontier of science, where molecular phylogeneticists started utilizing it as a prime tool to find out and establish the ancestral lineages of the predecessors in prehistoric or historic time and their contemporary descendants. Regarding the investigations related to mtDNA, it needs to be mentioned here that mitochondria or chondrisomes are the cellular organelles, which is found in the cell but found to be outside of the nucleus, which has the unique potential of escaping the "shuffling of genes occurs between generations" so the genes could have been transcended from mother to offspring unaltered through this cell organelle (Wade, May 2, 2000). So theoretically, for any descendant of the modern human individual should have the unaltered copy of DNA configurations in their mitochondria, they inherited from their mother. In reality, it hs been observed by the scientists that mtDNA had accumulated the changes of DNA sequences over centuries due to copy errors and mutagenic radiation-related changes due to sustaining exposition and adaptive interaction of the modern human species in the changing environments in its ambiance. However, after going through the primary data on the DNA sequence related investigations of molecular phylogeneticists, the molecular biologists have delved in the appraisal of size and dynamics of ancestral populations of the extant, derived representatives of that archaic humans and their tentative time of evolutionary splitting from the evolutionary cousins (lineages) in course of time. So, it is the matter of fact that emergence of DNA studies might have started to identify the individuals to find out its close phylogenetic associate, but the advancement of technological and perceptional progress in this frontier arena of science ushered its shift in paradigm to get involve the evolutionary scientists to carry on their quest for drawing through the dotted lines to explore the collective identity of the modern human race, found in between the prehistoric ancestors in the past and the contemporary descendants in this regime of Anthropocene.

In 1987, a Californian research team, engaged in the arena evolutionary genetics, lead by Allan C. Wilson, Rebecca Cann, and Mark Stoneking have used mitochondrial DNA (randomly collected from 147 individuals, belonging to 5 major geographical populations), tiny circular DNA rings, which has been attributed by the egg cells and transmitted through maternal lineage for constructing human family tree and this research trio have published their critical findings to the journal Nature under the title "*Mitochondrial DNA and human evolution*" (Cann et al, 1987). The preliminary result of this experiment has revealed that there are two distinct mtDNA lineages in that human population, the first human lineage has contained African mtDNA, whereas the second mtDNA one is of composite population type (different type of populations, except non-African type). The scientists have observed furthermore that the branch with African mtDNA type has been found close to the root of the tree. Based on the critical investigations of this primary data, the research investigators have been able to trace back the evolutionary divergence of the modern human species out of the common ancestral root-stock, likely to be lived between 99,000 to 148,000 years ago in Africa (Cann et al, 1987; aprilholloway, 29th April 2017). In the contemporary work, published in the journal, "*Cell*," entitled "*Mitochondrial DNA variation in human radiation and disease*" of Douglas C. Wallace (2015), from Emory University School of Medicine, Atlanta, USA, has briefly narrated that the high rate of mutation and maternal inheritance of human mtDNA is resulting consecutive accumulation of mtDNA genetic variants along with radiation of maternal lineages, that forms the mutational tree, which is comprised of "clusters of related mtDNA haplotypes," which has also been known as haplogroups. Wallace (2015) has critically reviewed the previous works of Cann et al, (1987); Kivisild et al (2006) and Wallace (2015) and has drawn an outline of human mtDNA phylogeny and geographic distribution based on a reconstruction of the origins and ancient migrations of women, that helped to contemplate that all African mtDNAs together (like L0, L1 and L2) to be considered as macrohaplogroup Land help to reconstruct a coalesce of a single origin of this group happens around 130,000 to 170,000 YBP (years before past). For 100,000 YBP, radiations and regrouping have found to be assumed among L haplogroups L0 mt DNAs is found to be the most ancient one, whereas L1 is found in Koi-san population and L2 in pygmy populations. Around 45,000 to 65,000 YBP, the L3 mtDNA has supposedly give rise to macro-haplogroups M (has contemplated giving rise to Asians and indigenous Americans Asia) and N macrohaplogroup (that has given rise to indigenous Europeans and Asians). Around 20,000 YBP, the haplogroups A, C, and D are confined to Northern Siberia, who have migrated through Bering Land Bridge to reach to form American indigenous people around 20,000 YBP and the B haplogroups have found to be settled in Pacific islands (MITOMAP, 2015; Wallace, 2015). Douglas Wallace had observed that almost all indigenous American Indians possessed mtDNA in their chondrisomes, that belonged to the lineages of A, B, C and D; whereas he recognized Europeans belong to two sub-sets: H, I, J and lineages and T, U, V, W and X lineages (Wade, May 2, 2000). The apparent gap or split between two sub-sets of a group of lineages in European tree, has rendered Wallace to contemplate that tentative arrival of modern human species to Europe happens around 39,000 to 51,000 YBP and he tried to reconcile the archaeological date of arrival modern human species to Europe, supposedly taken place around 35,000 years ago (Wade, May 2, 2000). According to the findings of Douglas Wallace, ancestral lineage M has been found in Asia with its descendants E, F, and G and he also noticed the presence of lineages A, B, C, and D in the Americas (Wade, May 2, 2000). Furthermore, Douglas Wallace had noticed that L lineage has found to be predominantly dispersed over throughout Africa with its three ancestral sub-sets like L0, L1, and L2; whereas L3 had contended to be the youngest lineage found in East Africa, which had contemplated to be the offset of Asian and European lineages. The contemporary reviews in the field of molecular phylogenetics, had regarded the academic observation and interpretation of mitochondrial lineages, propounded by Douglas C. Wallace as "haplogroups" but informally recognized as "daughters of Eve," as they have contended that all these lineages had branched out of the mother-stock of mitochondrial Eve in course of the evolutionary history of humanity. In this regard, an important breakthrough needs to be mentioned by population geneticists in 1999-2000 as they have observed that the size of the archaic human population was found to be confined in 2000 breeding individuals though the cross-investigating research reviews, done by another group of researchers, based on the data on the investigations on human mtDNA, has claimed that it failed to trace back 1000 women (having breeding potential) in such archaic population and the tree, based on mtDNA has been rooted in a sole individual, popularly recognized as "mitochondrial Eve" as the possible lineages might have been pushed into extinction, on its unsuccessful attempt to sustain and traverse through the evolutionary journey of humanity (Wade, May 2, 2000).

Instead of substantial progress in the molecular investigation concerning the evolutionary relationships of modern human species by analyzing the mitochondrial and autosomal DNA, it is difficult to understand the erstwhile human migration due to overlapping issues like inconsistent gene flow, genetic recombination, the random admixture of genes, random nature of mutations (specifically the convergent mutations) and enigmatic influence of natural selections, etc (Bowcock et al, 1991; Stoneking 1993; Shriver et al, 1993; Harding et al, 1997). To carry out a reconstruction of ancestral state of any population to define further on the state of polymorphisms, intermittent accumulation of mutation, historical mode of migration, etc. and deciphering its haplotype information that directs the history of humanity on Earth, scientists have engaged in the investigation on biallelic polymorphisms (which is integrated to the non-recombining segment of the Y chromosome (Jobling and Tyler-Smith, 1995; Hammer and Zegura, 1996; Underhill et al, 1997; Seielstad et al, 1999). In 1997, Peter A. Underhill, Peter J. Oefner along with seven research associates of Department of Genetics, Stanford University, USA, had successfully constructed a detailed family tree to elucidate the evolutionary history of modern human species, where integrated issues like genetic drift, historical migration, convergent nature of migration, and expansion of its range of contemporary distribution have been addressed by making a successful portrayal of Y chromosomal haplotypes based on analysis of Y chromosome of men. Based on the available data of the molecular investigation of the Y chromosomal DNA tree, entitled "*Detection of numerous Y chromosome Biallelic polymorphisms by denaturing High-Performance Liquid Chromatography*," the population geneticists had contemplated that archaic human population would have substantially small in size (Wade, May 2, 2000). Furthermore, the scientists worked on this Y chromosome tree project, had shared their observation that assumes that in each generation several ancestral men might not have any children or they might have only daughters rather than sons, that might have rendered gradual diminishing the Y chromosomes in the existing gene pool of the concerned population, even might have been a possibility when the population size remained unchanged (Wade, May 2, 2000). In 1997, L. Luca Cavalli-Sforza, the colleague of Peter Underhill and Peter Oefner has published an overview on this Y-chromosomal model in a review article, entitled "*Genes, People and languages*" in the journal "*Proceedings National Academy of Science, USA*," where he has observed that the tree has been rooted in a single trunk, Y chromosomal Adam and ten branches have been emerged out of this main trunk, have further been contemplated as 10 sons of Adam.

Fernando L. Mendez, one of the eminent human geneticist of Stanford University, USA, leading a research project, carried out along with 10 research associates, who had shared his unique discovery, made at the time of reviewing the phylogenetic reconstruction of an African-American Y chromosome, that possessed a large number of SNPs in the basal part of the Y chromosome (Mendez et al, 2013). Mendez and his research colleagues have successfully sequenced the 240kb sequence of this ancestral Y chromosome to identify the specific mutation on this lineage, and recognized that mutation as A00. Furthermore, team Mendez had able to determine the tentative time of the most recent common ancestor or (TMRCA), as well as the earliest known human ancestor of the modern human species, historically lived between 237,000 to 581,000 years ago. It has created a buzz in the field of phylogenetic researches to determine the tentative time of evolution of modern human species, out of its predecessor as well as the evolutionary age of TMRCA, that supposedly gave rise to the modern human race as it exceeds all available results of the evolutionary age of TMRCA, determined earlier by the scientists based on the genealogical investigations on mt DNA, and Y-chromosomal DNA and paleontological and anatomical investigations (age-dating examination) of the ancient human fossils (Mendez et al, 2013). Based on the critical review of the available results done by Mendez and his colleagues, the scientists have contended that A00 mutation had only been found in this rare type of ancestral lineage, possessed by the small size of the archaic human population, found to be found in low frequency in Central Africa. It is quite surprising to note that the scientists have traced out A00 mutation lineage among the data samples, collected from the African-American in origin, and these population had reportedly been descended from the "hunter-gatherer populations" from sub-Saharan Africa (Mendez et al, 2013). The genealogical investigation on this Y-chromosomal DNA of archaic human genome has rendered the scientists to make some important projections on the evolutionary history of modern human that an archaic introgression of the Y chromosome from ancient human populations most likely to be taken place in the genome of anatomically modern human species and the emergence of anatomically modern human species (and the emergence of Y-chromosome Adam) had supposedly taken place in Africa a long time ago, much earlier, than projected by the mtDNA discoveries (the evolutionary age of mitochondrial Eve) (Mendez et al, 2013). That indirectly indicates that the literary chance of genetic Adam and Genetic Eve to live together is less likely too.

The depiction of the evolutionary journey of humanity and reconciling the biblical narrative emergence of human on Earth is found to be most difficult to put together as the ideas and interpretation of an evolutionary interpretation of science about human species (like its gradual course of evolutionary emergence from ancestral root-stock of Great Apes) has not fit in the theological interpretation of divine creation of humanity (according to the God's grace), yet several earnest efforts have been made by some contemporary researchers, and the recently published article like "*Twelve degrees of separation: Do human haplogroups lead us to Adam or perhaps the Adityas*" had found to address several reconciling issues and logical concerns, needs to be reviewed with the open mindset (Keith, June 13, 2018). Before going into detail of this article, it requires to explain about the enigmatic phrase of "Twelve degrees separation," that originally comes from the notion of a theory "Six-degree separation," that contends that we, the modern human species are all connected by an invisible chain of acquaintances and any individual is just six introductions away from another person on this planet Earth. The theory of this average social distance about logarithmic relation to the population size has originally developed and published in "*Chain Links*" by Frigyes Karinthy, the eminent Hungarian playwright, journalist, and poet in 1929. This principle of social distance about its population size, prevailed in the modern human race sounds to be illogical in the world of 6.6 billion people but the experimental effort had found successfully validated it. Recently, a group of researchers from the Research and Development department of Microsoft Inc. The USA has critically reviewed the 30 billion electronic communications of 180 million people in different countries, dispersed all over the world (Smith, August 3, 2008; https://www.theguardian.com/technology/2008/aug/03/internet.email). The researchers have tried to examine the database to determine the minimum chain lengths that connect different pairs of 180 billion pairs and the average length of steps has found to be 6.6 to connect between two individuals. More precisely, the researchers have contended that 78% of pairs could have been connected in 7 hops, though in some extreme cases it might have taken up to 29 steps. In reality, the research project in-charge of Microsoft Inc. Eric Horovitz and his teammates firmly claimed that "*via the lens provided on the world by Messenger [MSN Messenger], we find that there are about 'seven degrees of separation' among people*" (Smith, August 3, 2008; https://www.theguardian.com/technology/ 2008/aug/03/ internet.email). So, in this regime of Anthropocene, when the modern human species had emerged as a diversified mega-population on planet Earth, in the norms of social distance, if a pair of human individuals is 7 degrees of separation from each other, in the historical time of biblical regime, when the human population has not assumed to be in such well-diversified state, 12 degrees separation is a feasible possibility, rather than be considered a myth. According to the biblical narrative of the emergence of ancestors of the modern human race on Earth had been started with the event of the special creation of Adam in the garden of Eden, which is assumed to happen in the dawn of human civilization, considered to be an act of special creation of an omnipotent creator (theologically recognized as God). As the biblical story has rolled on to the next level from the emergence of Adam in the dawn of civilization to the in the regime of Noah (who has assumed to be born in 3300 BC) as well as Genesis Flood, which has assumed to drown all living creations, once created by God, except 12 specific individuals (including Noah, his three sons, his three daughters-in-law) in the great Ark to survive such short span of mega-extinction to ensure that the depleting population of the human race could bounce back in post-flood regime to flourish and emerge as a diversified, modern human race found across almost every inhabitable niche on Earth (Radford, September 14, 2000; Keith, June 13, 2018). At this point, it needs to be mentioned that the geneticists and evolutionary scientists whoever believe and strongly support the theory of Intellectual design or Young Earth Creation, strongly argued such occurrence of Genesis Flood act, an event of a genetic bottleneck, that the ancestral human population has supposedly gone through before surging of the modern human population having taken place in the post-flood regime. So, following the biblical depiction of the journey of humanity, it has been found that from the very beginning of the journey of humanity that started with Adam to the first genetic bottleneck of Genesis Flood as well as regime of Noah, there are eleven generations (according to biblical perspective) of inbetween down the same line of inheritance like Adam, Cain, Abel, Seth, Enosh, Cainan, Mahalaleel, Jared, Enoch, Methuselah, and Lamech and finally it ends with Noah before we go to the next part of the story of Genesis flood where biblical narrative depicted the presence of 12 specific individuals, who took refuge to Noah's ark and the journey of the modern human race would have been started from there (Keith, June 13, 2018). In this context, it needs to be mentioned that the name of Cain, the elder son of Adam had always been mentioned with Abel, the younger son of him. According to the biblical narrative, by divine attribution of God, all these 12 individual super-humans lived a long period on Earth. According to the theological perspective, 12 is the magical figure, considered by several researchers, involved in deciphering the emergence of humanity on Earth as some researchers have referred the Vedic perspective of the genesis of humanity as several 12 Adityas (Sun-God) or 12 holy men has been mentioned in "*Vishnu Purana*," who had contemplated to be born out of the goddess Aditi and her husband holy sage Kayyapa, who had been stated as the son of Marichi and his wife Kala (Keith, June 13, 2018). According to "Vishnu Purana" and "*Srimad Bhagavatam Canto 12, Chapter 11, text:27-49*," another epic of Hindu religion, Marichi has been stated as the son of Brahma, the creator of universe and Earth, and these 12 holy Adityas have found to be responsible for beginning the journey of humanity on Earth and also form the foundation of early human civilization (Keith, June 13, 2018; Dalal, 2014).

A number research investigations pertaining the emergence of humanity on Earth have urged to reconcile the theological perspective of the historical emergence of the human race by erstwhile 12 super-humans as well as 12 human predecessors (across the Abrahamic and Hindu religious beliefs and practices) to the contemporary paleoanthropological and evolutionary interpretation of historical migration 12 haplogroups or early human family lineages, initiated the beginning of human civilization on this planet Earth, that begins with the maiden haplogroup lineage A (usually each capital letter like "A" onwards stands for a specific genetic line) and it has been continued in alphabetical order from A through L and split out from its sibling haplogroup lineage K (Keith, June 13, 2018). Several evolutionary researchers and the paleoanthropologists have noticed that such historical migration of ancient human populations and erstwhile migration of different haplogroup lineages that were witnessed in the historical actions of migrating clans of genetic family groups, started developing diverse cultural entities in different parts of the world and giving birth to new civilizations, matches found to be fitting in the socio-religious narratives of ancient human populations following Hebrew and Hindu traditional way of living and socio-religious practices (Thomas et al, 2002; Tofanelli et al, 2014; Keith, June 13, 2018). Those paleoanthropologists and the evolutionary scientists have further contended that in accordance to the Hebrew and Hindu traditional narratives, the name of these 12 haplogroups might be spelled differently, but the inherent identity of the erstwhile human lineages found to be identical and reconciled the scientific observation of the contemporary researchers observed on the migratory human genetic family groupings of haplogroups on this planet Earth, that existed and expanded exponentially to form this well-diversified human civilization of modern human species in this universe. Going forward, the paleoanthropologists and the evolutionists have contained that most of the haplogroup evolutionary lineages emerged independently as they represent a specific genetic line and follow a specific migratory track with exception of A and B as both lineages emerged as maiden lineages to initiate the evolutionary journey of humanity and stored all genetic information for all mutations that the derived group would have gone through over time and as a sibling and A emerged first followed by BT, that rendered emergence of B and further mutating (M168) to give rise to "Out of Africa" CT Haplolineage (1000 Genomes Project Consortium, 2010; Cruciani et al, 2011). The evolutionists have contended that the derived haploid-lineages, those that have emerged after B, have found to be successful in storing the genetic evidence of mutations taken place around 65,000 years ago (Keith, June 13, 2018). The researchers, engaged in depicting the historical migration of ancient human lineages over time reconciled such overlapping nature of migration, undertaken by A and B haplogroups found to be a metaphoric resemblance to the biblical narrative of Cain and Abel, when the early history of humanity has often been narrated that began with Adam and continued till Genesis flood where the population of ancestral human population restricted to 12, who took refuge to Noah's Ark to sustain such mega calamity on Earth (Keith, June 13, 2018). However, the paleoanthropological investigations on the historical migration of haplo-lineages across the globe, have found out that linage B had supposedly emerged around 80,000 years ago in Central Africa and dispersed over the continent, and carried the genetic information of haplo-lineage A along with it. It is quite intriguing to notice that the lineage E, which is represented by Bantu African aboriginal population, belonging to Pygmy tribes, originally descended from haplo-lineage B and lived in the tropical rainforest habitat of Congo and lead a traditional African-aboriginal life (Salas et al, 2002; Gomes et al, 2010; Keith, June 13, 2018). The anthropological investigations have ascertained that haplo-lineage has found to be endemic in the African continent as it had not diverged out of this continent, whereas the haplo-lineage C that has been reportedly emerged out of Africa around 60,000 Ybp, reportedly migrated to the Australian continent to become an aboriginal clan and a fraction of lineage C have moved further along the Arabian and Indo-Malayan region and initiated new civilizations in the wide peninsular regions in the New world (Karachanak et al, 2013; Bergstrom et al, 2016; Keith, June 13, 2018). Furthermore, the researchers shared their critical investigations on the genealogical evidence of the historical migration of the haplo-lineage C across the globe, revealed that the genetic foot-prints C have been found in its descendants, migrated away far and wide and represented by the contemporary human populations in middle-East Asia, Indian subcontinent and South-East Asia (Karachnak et al, 2013; Keith, June 13, 2018).

The paleoanthropological investigations have ascertained that the haplo-lineage D had migrated to far-East like Japan and Andaman Islands, those had assumed to be connected to the mainland of Asia in those early days of continental drifting and a fraction of D had supposedly reached Tibetan plateau, eventually in search of a habitable niche for ancestral human populations in historical time (Hammer et al, 2006). So, it rendered the anthropologists to contemplate that a successful attempt of human migration of a human population, belonging to a particular clad (D), might have given rise to three distinct human entities, having its distinctive geographical and linguistic uniqueness in the backdrop of geographical and environmental isolation (Hammer et al, 2006; Shi et al, 2008). The anthropologists are mesmerized with the research investigations on the antiquity of the erstwhile D lineage, when they came across the comparative studies on the physical and cultural anthropological similarities and uniqueness of the contemporary population of the indigenous human population belonging to Aino in Japan, indigenous people of Tibet and indigenous people of Andamans, proved to be descended from the ancestral haplo-lineage D in historic time (Shi et al, 2008; Keith, June 13 , 2018). Haplo-lineage E is found to be the evolutionary sibling of D, which has reportedly migrated towards the southern part of the African continent and gradually occupied a vast region of the African continent, once occupied with the haplo-lineage A and B and the anthropologists have referred the analogical resemblance to this incident to the migration-driven settlements in 1500 BCE, by Aryans, nomadic migrated from Central Asia, crossed over the Hindu Kush Mountains and migrated to Indus Valley area (one of the ancient farming based human civilization, founded around Indus river flood plain in Indian subcontinent around 5,500 BCE) to settle down there that rendered changes of the nomadic lifestyle of a tribe to embrace agrarian community-based living by them as a result of a successful attempt of migration-driven settlement (Semino et al, 2004; Keith, June 13, 2018).

The anthropologists have observed that haplogroup F would have given rise to the nest of derived lineages GHIJK, around 50,000 years ago and the human populations, specifically the men descended from such derived lineages, supposedly genetically characterized by a mutation footprint M89 (Semino et al, 2004; Keith, June 13, 2018). In this point, the anthropologists have observed the emergence of caste systems, that has segregated the homogenous human entities to certain social groups, called castes based on professional skills and distinction of social-professional role-play, that society needs at that time to go for collective progress and maintaining social harmony for a civilization to form and progress. According to the traditional Vedic-Hindu narratives, God had created four caste systems in human communities, out of his body and that socio-religious emergence of four castes had supposedly taken place between 60 to 65 millennia ago (Keith, June 13, 2018). However, the anthropologists have contended the existence of the nested haplo-lineages as well as recognized GHIJK as conglomerate clades, which has grown out of haplo-lineage F and recognized the "third diffusion of four castes into society" in the passage of 10,000 of years and the haplo-lineage G has been recognized as a most competent farming community, whereas the haplo-lineage J (the Semitic community) has assumed to emerge as experts in Commerce sector, the haplogroup I (the ancestral European communities) has emerged as experts in the ancient industrial arena (Dato et al, 2004; Rollo et al, 2006; Keith, June 13, 2018). The anthropological reviews have revealed further that haplo-lineages I and J had split out from K, when the emergence of L and T had taken place at the same time and the genetic footprints of haplo-lineage L have been witnessed among the contemporary dispersed human communities, inhabited in Malabar coasts in Southern India in Indus delta and the mountains of the Indian subcontinent, erstwhile heartland of Harappan civilization (Houshmand et al, 2005; Rollo et al, 2006; Keith, June 13, 2018). Furthermore, the anthropologists have observed that over time, haplo-lineage K has found to be disintegrated, as it left the GHIJK conglomerate clades to trigger its siblings to get in secondary migration to spread over all terrestrial landmasses, all over the world, in quest of occupying even the unexplored landscape in the remote corner on Earth (Rollo et al, 2006).

However, regarding the historical depiction of migration of ancient human populations, it needs to be mentioned that according to the biblical narrative, Adam and Eve had three sons Cain, Abel, and Seth, but the contemporary findings based on the DNA data have rendered the molecular geneticists to come up with a new proposition where 10 sons were descended from genetic Adam, whereas 18 daughters were found to be descended out of genetic Eve (Wade, May 2, 2000). However, out of which, the first three lineages of ten, represented by sons I, II, and III have found to be widespread in distribution but geographically restricted in the African continent. son III hi assumed (based on findings and logical interpretation) to be migrated to Asia, lineage IV or son IV has supposedly migrated to the Sea of Japan, lineage V or son V has assumed to be migrated to northern India and son VI and X construed to be migrated to Southern Caspian sea (Cavalli-Sforza,1997). Furthermore, the critical analysis of the investigation on Y chromosomal DNA of Underhill et al (1997), rendered Luca Cavalli-Sforza (1997) to postulate that all Y chromosomal lineages might have been associated to the major language groups all over the world as he has cited the example that the South-Caspian population might have spoken in Eurasian, which has also anthropologically recognized as the erstwhile tongue of Indo-European and other major language families of the continent but the identical finding of Wallace, based on mtDNA lineages made he a cautious observation as he stated "*…tended to be more cautious than Luca*" (Wade, May 2, 2000).

Whether its "Mitochondrial Eve" hypothesis of Douglas Wallace or "Y chromosomal Adam" hypothesis of Peter Underhill and his associates, both postulations have projected towards a common observation that the origin and evolution of ancestral lineages of modern human species happen in Africa as it has been stated with the affirmation "...geneticists believe, and started to split up some time after 144,000 years ago, give or take 10,000 years, the inferred time at which both the mitochondrial and Y chromosome trees make their first branches" (Wade, May 2, 2000). According to the special creation myth of Abrahamic religious perception, Adam and Eve have been considered as the primordial couple in the garden of Eden, as they have been considered to be the predecessor of the modern human race on Earth. Whereas the critics of Abrahamic religious followers and the atheists have refuted the idea of the special creation of human species as well as the supernatural creation of modern human race by the grace of God. The contemporary progress of scientific research in the field of molecular genetics has rendered the scientists to firmly stand behind their claim that origin and evolution of modern human species have descended out of one historical man and one historical woman, around 100,000 Ybp (aprilholloway, July 29, 2017). However, a new group of geneticists has started claiming themselves as the bonafide supporters of either "intelligence design" or ID hypothesis or "Young Earth Creation" or YEC hypothesis and they have come up with a spirit of an amicable reconciliation between the radical evolutionists and fundamental special creationists to narrate the history of humanity so they have started recognizing the primordial woman as genetic Eve or mitochondrial Eve and the primordial man as genetic Adam or Y chromosomal Adam. Some molecular investigations have been made since 1987 to establish the evolutionary identity of mitochondrial Eve and Y chromosomal Adam to estimate their tentative time of emergence and the estimation of the molecular clock has left a formidable gap as it shows that they had supposedly emerged between 60,000 to 580,000 Ybp (aprilholloway, July 29, 2017).

However, in 2013, G. David Poznik along with ten research colleagues from School of Medicine, Stanford Univerity, had published a research paper, entitled "*Sequencing Y chromosomes resolves discrepancy in time to common ancestor of males versus females*" in the journal of "*Science*" that claimed that every extant male member, belonging to modern human species, had descended from the ancestral male member belonging to human species lived 135,000 Ybp and this primordial male supposedly lived in the historic time whenever the primordial woman and the mother of all woman lived on Earth around the similar time the primordial man existed and they have popularly been recognized as Y-chromosomal Adam and Mitochondrial Eve as well as a patrilineal and matrilineal ancestor of the modern human race in the proud history of origin and evolution of humanity on Earth, based on the contemporary investigation on Y-chromosome DNA and Mitochondrial DNA (aprilholloway, July 29, 2017). However, G. David Poznik and his research colleagues (2013) have sequenced the genome of 69 males, belonging to 9 populations and identified the ancient phylogenetic structure in the African haplogroup. The scientists have applied same methodologies to the Y chromosome and mitochondrial DNA to estimate the time to the most recent common ancestor (MRCA) of the Y chromosome, between 120,000 to 156,000 Ybp, whereas the time to the most recent common ancestor (MRCA) of Mitochondrial DNA, is in between 99,000 to 148,000 Ybp (Poznik et al, 2013). Furthermore, the scientists had observed that all members in the global representation, shared a common ancestry with one single man from Africa and the male lineages have coalesced in significantly in recent times compared to female lineages. Such findings might have created excitement to the supporters of special creationists as it seems like that biblical narrative of the emergence of humanity on Earth but Poznik and his team members have made it clear based on their observation that Adam and Eve of the Bible are not the identically same to genetic Adam and Eve and they have stated further "They weren't the first modern humans on the planet…but instead just two out of thousands of people alive at the time with an unbroken male or female lineages that continue on today" (aprilholloway, July 29, 2017). To make things clear, Poznik and co-researchers have observed that the genetic Adam and Eve might not have had a chance to live together to reproduce and the date of emergence, estimated experimentally for genetic Adam and Eve, seems to be a rough estimation of the tentative evolutionary age of our MRCA depends upon the diversities of DNA sample collected from a diverse range of human populations, found to be representing the global population of modern human species, we need to choose carefully to carry out our DNA sequencing investigations to yield a precise result (aprilholloway, July 29, 2017).

Since discoveries of Mitochondrial Eve and Y chromosome Adam, these two names mesmerized our collective imagination that almost everyone including scientific and non-scientific communities started referring this primordial couple as our common maternal and paternal ancestors. It is found to be a non-scientific approach as it yields an over-simplified perception that we inherit DNA, vertically descended from this couple through a series of intermediate generations so deciphering the evolutionary history of humans might have been explored by undertaking the molecular investigations of mitochondrial DNA and Y chromosomal DNA. In 2005, National Geographic Society has come forward to turn such possibility to reality as it launched a world-wide DNA data mining project to collect and analyze data of mtDNA and Y chromosome DNA from a million people, belonging to diverse ethnic populations and subpopulations all over the world and this project has popularly been recognized as "genographic project" (Reich, April 8, 2019). At the initial phase, it was found by the scientists, working in this Genographic Project, that this project seems to be outdated as most of the relevant data had already been mined and the primary analysis of these investigations have found to be quite interesting. Furthermore, the scientists of this genographic project have contended that the untold history of origin and evolution of humanity, encrypted in the mitochondrial DNA and Y chromosome data have mostly been mined and decrypted and that effort has not sufficed to go to make an earnest effort to decipher the buried evolutionary history of humanity from its genome without exploring the entire genome, including the non-coding regions of it (Reich, April 8, 2019). In their in-depth genealogical exploration in the biological holy grail of modern human species, has rendered the scientists to contemplate that human genome is not a simple collection of two stories that revealed the emergence of two distinct genealogical lineages, traced back through the mitochondrial DNA and Y chromosome of the human genome, rather it's a treatise of a big collection of short stories of our diverse group of predecessors, which reveal the existence of thousands of "independent genealogical lineages," traced back to the genome of archaic human populations (Reich, April 8, 2019). The molecular genomicists have tendered a unique analogy to convince the functional and structural aspects of DNA sequences, arrayed in the human genome, to define it as a matrix of 46 mosaic tiles (i.e. chromosomes) rather than a continuous, linear stretches of DNA sequence, traced back from a single ancestor and it's found to be like that because the long segments of DNA have found to be segmented as well as separated inside the cell. We are very much aware of our primary structure of human genome that in a genome there are 23 chromosomes and each human individual possesses two genomes (one of it inherited from paternal lineage and another one from maternal lineage) so the total number of chromosomes in any individual of modern human species is found to be around 46, that eventually form a matrix of mosaic tiles (if and when we consider each chromosome as a mosaic tile) (Reich, April 8, 2019). The scientists have tried to present their observation by a simple narrative of such matrix of mosaics, formed by the chromosomal arrangements in its nuclear region of the cell, as they have cited a convincible hypothetical scenario that one-third of the chromosomes of a woman that passed down to her egg, might have been inherited through paternal lineage, whereas the rest of the two-third of her chromosomes inherited from maternal lineage and the parental copies of chromosomal mosaics, splicing together, have been possessed in the ovary of the concerned woman. After critical reviewing of such splicing pattern, the molecular genomicists have come up with their observations as they contended that on average 71 splices have been produced per generation of modern human species, while the males yield 26 splices during spermatogenesis (formation of sperms) and females yield 45 splices during oogenesis (formation of an egg) (Reich, April 8, 2019). That helps us to trace back each generation to its past, we could easily assume that the genome of the person concerned, has genetically descended from an increased number of spliced copies of ancestral fragments, so it rendered to contemplate further that the genome of the present human population as well as in our possession mirrors the ancestral diversities in our past (Reich, April 8, 2019). The genome biologists have ascertained that any person's genome has been derived from 47 stretches of DNA (out of 47, 46 stretches have been inherited from both parents and 1 mitochondrial DNA from mother) so if we would be able to do simple math we would see how the DNA stretches would have been transmitted vertically from the previous generation (immediate) of the present individual, as it shows:

i. In the first generation back the genome of the person concerned has supposedly inherited (47+71) or 118 stretches of DNA from its parents

ii. Likewise in the second generation back, the genome of the person concerned has inherited (118+71) or 189 stretches of DNA from its four grandparents

So, the statistical calculation has revealed that increasing number of ancestral stretches of DNA in every generation has overhauled by the two-fold size of the ancestral population and a tentative extension of this experimental simulation to go back to tenth generation cycle, it has been observed that the number of ancestral stretches of DNA has found to be 757, where the size of the ancestral population has supposedly found to be around 1.024, indicating that person has supposedly had a few hundred ancestors, from whom the person concerned had hardly been inherited with any stretches of DNA (Reich, April 8, 2019). So, if it would have been extended further backward, in the twenty generations back, the number of ancestors are around thousand times larger than the number of ancestral stretches of DNA, found in the genome of the person concerned, so it would empirically project that every person would have not derived the ancestral stretches of DNA from his and her ancestral lineages. Hence, it is a proven fact that the genealogical reconstruction of any person to determine its evolutionary lineage, based on statistical or mathematical calculations, from its historical record, might not reconcile his and her actual genetic inheritance (Reich, April 8, 2019). Furthermore, the molecular genomicists have strived to traverse through "deeper in time" and realized that the appearance of the genome of a person looking scattered with ancestral fragments of DNA, descended from a fairly large number of the ancestral population. So, an evolutionary stretch back to fifty thousand years earlier from present, the genome of the modern human species has conceivably scattered in hundreds of thousand ancestral fragments of DNA and that has appeared to be much bigger than the number of individuals, belonged to that erstwhile population in that historical time and that scenario rendered the population biologists, molecular biologists, and evolutionary biologists that the derived group of contemporary human populations and subpopulations, widespread all over the world, most likely derived DNA from almost each and everyone from the ancestral population, who supposedly had a large number of offsprings in the remote past (Reich, April 8, 2019).

However, to substantiate such shortcomings of genealogical reconstruction concerning the determination of evolutionary lineages, several population biologists and genomicists have referred the chronicles of the family tree of Royal families in great Britain, which has carefully been prepared and maintained for ages, where the genealogical data has indicated that that Queen Elizabeth II of England unlikely derived any DNA from William of Normandy, the erstwhile conqueror of England in historical time and assumed to be the predecessor of Queen Elizabeth II around 24 generations back (Reich, April 8, 2019). At the same time, the scientists have not blatantly ruled out any possibility of such inheritance of ancestral stretches of DNA by Queen Elizabeth II from William of Normandy as they have stated: "*This does not mean that Queen Elizabeth II did not inherit DNA from ancestors that far back, just that it is expected that only about 1,751 of her 16,777,216 twenty-four-degree genealogical ancestors contributed any DNA to her*" (Reich, April 8, 2019). Technically, the geneticists have considered such possibility as a remotely plausible incident in the arena of the evolutionary history of humanity as the result is empirically such obscuring in reality that considering William as Queen's genetic ancestor among the thousands of possible genealogical lineages seems to be logically untenable, if and even a high level of occurrence of inbreeding would have been considered in the Royals of Great Britain to substantiate such remote possibilities.

Let's go back to square one, when the narratives of the special creation of emergence of humanity on Earth sounds inconvincible to a group of radical evolutionists, the scientific narratives of an evolutionary depiction of humanity, supported by genealogical evidence and interpretations sound like an incomplete portrayal of a picture to theologists or someone, engaged in religious practices, where it requires a reconciliation between these two to yield the best outcome. Either from the divine perspective of interpretation of the emergence of life or evolutionary interpretation of the emergence of life on Earth, tried to define whether presence and existence of modern human species have expected to be inevitable to most of us or not, it sounds imprudent to most of us as it does not matter whether we are religious people or followers of scientific principles or even atheists, our first and foremost dream to see ourselves on the top of pyramids of all ecosystems should be considered as a collective endeavor rather than an individual aspiration. In this age of digital as well as virtual era, if we prioretize importance of science over religion and theoretically consider evolution as driving impetus and potential over special design, to create an entity and ensure its sustainability, cosmic evolution might have created emergence of this universe from cosmic fusion of clouds, vortex of gases gases, galactic dusts, but without emergence of the planet Earth, emergence of universe would have remained an incomplete story, likewise, the inorganic and organic evolution as well as emergence of planet Earth (as blue planet) from its erstwhile state of dark state of proto-Earth, has found to remain incomplete, if the emergence of life would not have been there; likewise the biological evolution of life of Earth and its further extention of emergence of modern human species on Earth has found to be an integral part, rendering completion of its existence as a structural and functional (ecological) entity of it and the universe it belonged to by maintaining an equilibrium state of living and non-living entities in our ambiance, a number of people have acknowledged it as intellectual design. Yet, the possibility of exploration of divine interpretation of scientific principles or vice versa would have remained a hypothetical scenario in this regime of Anthropocene, that would have been left to our future generations to find out.

References

Amos, W., and Hoffman, J. I. (2010). Evidence that two main bottleneck events shaped modern human genetic diversity. *Proc Biol Sci.* 277(1678): 131-7. doi: 10.1098/rspb.2009.1473.

aprilholloway, (April 29, 2017). Genetic "Adam and Eve": All humans are descendants of one man and woman who lived over 100,000 years ago. https://www.ancient-origins.net/news-evolution-human-origins/genetic-adam-and-eve-all-humans-are-descendants-one-man-and-woman-who-021536

Bergström, A., Nagle, N., Chen, Y., McCarthy, S., Pollard, M. O., Ayub, Q., Wilcox, S., Wilcox, L., van Oorschot, R.A., McAllister, P., Williams, L., Xue, Y., Mitchell, R. J., Tyler-Smith, C. (2016). Deep Roots for Aboriginal Australian Y Chromosomes. *Curr Biol.* 26(6): 809-13. doi: 10.1016/j.cub. 2016.01.028.

Bowcock, A.M., Kidd, J.R., Mountain, J.L., Hebert, J.M., Carotenuto, L., Kidd K. K., and Cavalli-Sforza, L.L. (1991). Drift, admixture, and selection in human evolution: a study with DNA polymorphisms. *PNAS U S A.* 88(3):839-43. doi: 10.1073/pnas.88.3.839.

Cann, R., Stoneking, M. and Wilson, A. (1987). Mitochondrial DNA and human evolution. *Nature* 325: 31–36. https://doi.org/10.1038/325031a0

Cavalli-Sforza, L. (1997). Genes, peoples, and languages. PNAS 94 (15) 7719-7724; https://doi.org/10.1073/pnas.94.15.7719

Cruciani F, Trombetta B, Massaia A, Destro-Bisol G, Sellitto D, Scozzari R. (2011). A revised root for the human Y chromosomal phylogenetic tree: The origin of patrilineal diversity in Africa. *Am J Hum Genet* 88: 814–818.

Dalal, R. (2014). *Hinduism: An alphabetical guide*. Penguin, UK. P.520.

Dato, S., Passarino, G., Rose, G., Altomare, K., Bellizzi, D., Mari, V., Feraco, E., Franceschi, C., and De Benedictis, G. (2004). Association of the mitochondrial DNA haplogroup J with longevity is population specific. *Eur J Hum Genet* 12:1080–1082. doi:10.1038/sj.ejhg.5201278

Gomes, V., Sánchez-Diz, P., Amorim, A., Carracedo, A., and Gusmão, L. (2010). Digging deeper into East African human Y chromosome lineages. *Hum Genet.* 127(5): 603-13. doi: 10.1007/s00439-010-0808-5.

Hammer, M. F., Karafet, T. M., Park, H., Omoto, K., Hirahara, S., Stoneking, M., and Horai, S. (2006). Dual origins of the Japanese: common ground for hunter-gatherer and farmer Y chromosomes. *Journal of Human Genetics* 51:47–58.

Hammer, M. F., and Zegura, S. L. (1996). The role of the Y Chromosome in human evolutionary studies. *Evol. Anthropol.* 5:116–134.

Harding, R. M., Fullerton, S. M., Griffiths, R. C., Bond, J., Cox, M. J., Schneider, J. A., Moulin, D. S., and Clegg, J. B. (1997). Archaic African and Asian lineages in the genetic ancestry of modern humans. *Am J Hum Genet.* 60(4): 772-89.

Houshmand, M., Sanati, M. H., Babrzadeh, F., Ardalan, A., Teimori, M., Vakilian, M., et al. (2005). Population screening for association of mitochondrial haplogroups BM, J, K and M with multiple sclerosis: interrelation between haplogroup J and MS in Persian patients. *Mult. Scler.* 11: 728–730. doi: 10.1191/1352458505ms1228sr

Jobling, M.A., and Tyler-Smith, C. (2003). The human Y chromosome: an evolutionary marker comes of age. *Nat Rev Genet.* 4(8): 598-612. doi: 10.1038/nrg1124.

Karachanak, S., Grugni, V., Fornarino, S., Nesheva, D., Al-Zahery, N., Battaglia, V., Carossa, V., Yordanov, Y., Torroni, A., Galabov, A.S., Toncheva, D. and Semino, O. (2013) Y-Chromosome Diversity in Modern Bulgarians: New Clues about Their Ancestry. *PLoS ONE* 8(3): e56779. https://doi.org/10.1371/journal.pone.0056779

Keith, S. (June 13, 2018). Twelve Degrees of Separation: Do Human Haplogroups Lead Us to Adam or Perhaps the Adityas?

https://www. ancient-origins.net/human-origins-science/twelve-degrees-human-haplogroups-adam-adityas-021940

Kivisild, T., Shen, P., Wall, D.P., Do, B., Sung, R., Davis, K., Passarino, G., Underhill, P. A., Scharfe, C., Torroni, A., Scozzari, R., Modiano, D., Coppa, A., de Knijff, P., Feldman, M., Cavalli-Sforza, L.L., and Oefner, P.J. (2006). The role of selection in the evolution of human mitochondrial genomes. Genetics. 172(1): 373-87. doi: 10.1534/genetics.105.043901.

McKie, R. (February 10, 2018). Cheddar man changes the way we think about our ancestors. (The Observer: Genetics) https://www.theguardian.com /science /2018/feb/10/cheddar-man-changed-way-we-think-about-ancestors

Mendez, F. L., Watkins, J.C., and Hammer, M. F. (2013). Neandertal Origin of Genetic Variation at the Cluster of *OAS* Immunity Genes, *Molecular Biology and Evolution*, 30(4): 798–801,

https://doi.org/ 10.1093 /molbev/ mst004

MITOMAP (2015). A Human Mitochondrial Genome Database. http://www. mitomap.org.

Poznik, G. D., Henn, B. M., Yee, M. C., Sliwerska, E., Euskirchen, G. M., Lin, A. A., Snyder, M., Quintana-Murci, L., Kidd, J. M., Underhill, P. A., and Bustamante, C. D. (2013). Sequencing Y chromosomes resolves discrepancy in time to common ancestor of males versus females. *Science.* 341(6145):562-5. doi: 10.1126/science.1237619.

Radford, T. (September 14, 2000). Evidence found on Noah's ark flood victim. https://www.theguardian.com/science/2000/sep/14/internationalnews.archa eology

Reich, D. (April 8, 2019). One hundred thousand Adams and Eves? https://www.ancient-origins.net/human-origins-science/mitochondrial-eve-0011717

Rollo, F., Ermini, L., Luciani, S., Marota, I., Olivieri, C., and Luiselli, D. (2006). Fine characterization of the Iceman's mtDNA haplogroup. *Am J Phys Anthropol.* 130(4): 557-64. doi: 10.1002/ajpa.20384.

Salas, A., Richards, M., De la Fe, T., Lareu, M.V., Sobrino, B., Sánchez-Diz, P., Macaulay, V., and Carracedo, A. (2002). The making of the African mtDNA landscape. *Am J Hum Genet.* 71(5): 1082-111. doi: 10.1086/344348.

Semino, O., Magri, C., Benuzzi, G., Lin, A.A., Al-Zahery, N., Battaglia, V., Maccioni, L., Triantaphyllidis, C., Shen, P., Oefner, P.J., Zhivotovsky, L.A., King, R., Torroni, A., Cavalli-Sforza, L. L., Underhill, P. A., and Santachiara-Benerecetti, A. S. (2004). Origin, diffusion, and differentiation of Y-chromosome haplogroups E and J: inferences on the neolithization of Europe and later migratory events in the Mediterranean area. *Am J Hum Genet.* 74(5): 1023-34. doi: 10.1086/386295.

Shi, H., Zhong, H., Peng, Y., Dong, Y. L., Qi, X. B., Zhang, F., Liu, L. F., Tan. S. J., Ma R. Z., Xiao, C. J., Wells, R. S., Jin, L., Su, B. (2008). Y chromosome evidence of earliest modern human settlement in East Asia and multiple origins of Tibetan and Japanese populations. *BMC Biol.* 6: 45-10.1186/1741-7007-6-45.

Shriver, M.D., Smith, M.W., Jin, L., Marcini, A., Akey, J. M., Deka, R., and Ferrell, R. E. (1997). Ethnic-affiliation estimation by use of population-specific DNA markers. *Am J Hum Genet.* 60(4) :957-64.

Seielstad, M., Bekele, D., Ibrahim, M., Touré, A., and Traoré, M. (1999). A view of modern human origins from Y chromosome microsatellite variation. *Genome Res* 9:558–567.

Smith, D. (August 3, 2008). Proof! Just six degrees of separation between us. (The Observer technology).

https://www.theguardian.com/technology /2008/aug/03/internet.email

Stoneking, M. (1993). DNA and recent human evolution. *Evolutionary Anthropology* 2(2): 60-73.

The 1000 Genomes Project Consortium. (2010). A map of human genome variation from population-scale sequencing. *Nature* 467: 1061–1073

https://doi. org/10.1038/nature09534

Thomas, M.G., Weale, M.E., Jones, A.L., Richards, M., Smith. A., Redhead, N., Torroni, A., Scozzari, R., Gratrix, F., Tarekegn, A., Wilson, J.F., Capelli, C., Bradman, N., Goldstein, D.B. (2002). Founding mothers of Jewish communities: geographically separated Jewish groups were independently founded by very few female ancestors. *Am J Hum Genet.* 70(6): 1411-20. doi: 10.1086/340609.

Tofanelli, S., Taglioli, L., Bertoncini, S., Francalacci, P., Klyosov, A., and Pagani, L. (2014). Mitochondrial and Y chromosome haplotype motifs as diagnostic markers of Jewish ancestry: a reconsideration. *Front Genet.*5:384. doi: 10.3389/fgene.2014.00384.

Underhill, P. A., Jin, L., Lin, A.A., Mehdi, S.Q., Jenkins, T., Vollrath, D., Davis, R.W., Cavalli-Sforza, L.L., and Oefner, P.J. (1997). Detection of numerous Y chromosome biallelic polymorphisms by denaturing high- performance liquid chromatography (DHPLC). *Genome Res.* 7:996– 1005.

Wade, N. (May 2, 2000). The human family tree: 10 Adams and 18 Eves. https://www.nytimes.com/2000/05/02/science/the-human-family-tree-10-adams-and-18-eves.html

Wallace, D. C. (2015). Mitochondrial DNA variation in human radiation and disease. *Cell.* 163(1):33-8. doi: 10.1016/j.cell.2015.08.067.

Chapter 8

Heterozygous Adam & Eve: Its relation to Population Constriction & Genetic Drift

Expressions of genetic drift in the current human genome will be considered together with the several other factors that have shaped that genome. More specifically, the involvement of introgression, niche construction, and selective breeding or domestication will be examined at length, and gene-culture co-evolution will be identified as the prime mover of recent human development. All of these explanations of evolutionary processes contrasting with plain natural selection are at significant variance with the currently still dominant paleoanthropological and archaeological model of recent human evolution, the replacement hypothesis ('African Eve') and the various other Out-of-Africa (OoA) models, which have governed most relevant discourse over the past few decades.

---- ***Robert G. Bednarik***

Evolution has an abstract form resemblance to "liquid" which fits in a various number of templates (containers) irrespective of the size or shape of it but the identity of any liquid fluid (as a distinct state of matter) remains universal, likewise, the functional interpretation of Evolution has been witnessed and interpreted distinctly by the experts, belonging to distinct inter-disciplinary branches of science, particularly in the broad arena of biology ranging from molecular biology, evolutionary biology, population biology and so on, but the cardinal aim and objective of evolution are to decipher and scientifically narrate the occurrence of changes of heritable features (those have been regulated by the expression of genes, encrypted in the genome of the concerned species of that population) of biological populations, that passed through generations to generations to render a species to attain a distinct form from its ancestral form. To a strong follower of the Darwinian doctrine of biological evolution, natural selection is the primary impetus that triggers the biological evolution in action. It creates an illusion among the supporter of evolutionists whether natural selection is the sole driving force that triggers and the wide-spread investigations have ascertained that natural selection is not the sole source of evolution as mutation and genetic drift play a major role in coordinating this non-directional accomplishment (beginning to end) that render a species to sustain in the Spatio-temporal dynamic environment on Earth. However, in the backdrop of population genetics, genetic drift has been recognized as a driving process of evolution in action by changing of the genetic profile, mainly due to chance in terms of its alteration of frequency of alleles (a version of a gene) of any species, belonging to a population over a generation, that happened primarily due to chance. So evolution depicts the change of allele frequencies of any population over generations, triggered by the composite action of natural selection, mutation, genetic recombination, and genetic drift, etc. to make the species better suited to the changing environment to sustain and pushed the unsuitable one to the brink of extinction. The geneticists have observed that genetic drift plays an important role in changing the allele frequencies of any population over generations, and it is a randomly driven impetus rather than a definite selection driven process, that puts evolution in action. So, if we consider genetic drift as a state of allele frequencies in a population from generation to generation, it intrigues us to comprehend how do we recognize it? If we try to find out the evolutionary functionality of genetic drift it's quite interesting as the genetic drift leads to loss of some alleles (even beneficial ones), followed by another cation called fixation, that ensures enhancement of the level of frequency of other alleles, and that renders the alteration of the phenotypic profile of the concerned species. The population biologists have defined and further observed that genetic drift has been appeared as "accidental random event" caused due to "sampling error" in course of selection of alleles from the gene pool of the present population to transcend to the future generation and the random effect of genetic drift has been witnessed across any size and shape of the biological population but the population biologists have observed that its impact on the small population is very intense (Bednarik, 2011). The population biologists have observed that the genetic drift has an immense effect on those small or fragmented populations (from the main populations), caused by the drastic impact of natural disaster, recognized as "bottleneck effect." The new small-sized break-way colony, that started acting founding of the new colony, has been recognized as "founder effect."

If we look around the beginning of genetic drift as an integral part of the evolutionary history, and initially it has found to be floated as a perception by Sewall Wright in 1929, emerged as an important genetic mechanism, made the evolutionary changes in action. In his "neutral theory of molecular evolution," Kimura (1968, 1983) had established the perception of "genetic drift" as the key functional factor that potentially steers the journey of biological evolution keep going and he also scientifically established that the major evolutionary changes, that we have witnessed at the molecular level, mainly driven by "stochastic process of the drift of selectively neutral mutants." So, next to Darwinian principles of "natural selection" (Darwin, 1859), genetic drift has found to hold as an evolutionary determinant that promotes the act of gene variation to attribute the reproductive success of any population of a species. Though natural selection and genetic drift, both mechanisms, are primarily responsible for changing the allele frequencies over time in course of its way of action, both are recognized as a distinct type of evolutionary mechanisms as genetic drift clearly shows that accumulation of alleles in offspring represent the random samples of their parents but population biologists ascertained that according to the probability theory, the impact of genetic drift on a large size of the population is negligible, even though a fraction of zygotes with allele frequencies, would have got chance to be matured. So, the next question that comes to mind is that as an evolutionary mechanism, genetic drift really has potential to make any substantial evolutionary changes that rendered an individual organism with altered allele frequency to survive or thrive to be diverged out as a distinct evolutionary form on Earth. The evolutionists and the population biologists have observed that unless the small population size had sustained the bottle-neck effect or founding effect, the offspring individuals (belonging to new population) would have gone through the course of genetic drift, but failed to produce adaptations most of the time. The reproductive biologists and geneticists have observed several possibilities, like an assortment of chromosomes during meiotic cell division, and due to linkages, etc., all parental alleles, would not ever be transcended from the sexually reproduced, diploid parents to their offsprings, particularly those animals having low reproductive numbers and longer reproductive cycles (Suzuki et al, 1989). The scientists had noticed that in small populations, genetic drift leads to loss of some genetic traits and emergence of some populations with unique genetic traits without any genetic or survival values, yet such populations sustained with the phenotypic expressions genetic drift of bottleneck effect and founder effect and found to be distributed over in some geographical locations. As an empirical reference of genetic drift, scientists have referred the example of genetic variability of "northern elephant seal," the population size of it had drastically shrunk to 20 in 1890 and the contemporary population size had recovered to 1500 times of that depleting size of the population under the present intensive conservation measure to offset the pressure of over-exploitation of it (excessive hunting) should be considered as an ideal example of genetic drift as the scientists have contemplated that the changes of genome size of an elephant seal in last 200 years has supposedly taken place due to drastic population shrinkage, rather than tendering hypothetical incompatibility to fit into natural selection.

The population geneticists most often refer to the occurrence of Icelandic cattle as a popular example of a founder effect, which is genetically related but distinct from the Norwegian subpopulation (Deme), it had diverged from. For human beings, scientists have referred the historic, insular, subpopulations of Palau and Flores as an ideal example of founder effect on the human species. The paleontological evidence of primates, found in the Flores islands of Upper Pleistocene period, has often been treated as an extinct hominin, recognized as *Homo floresiensis* by a group of scientists, whereas the critical level of investigation on its genetic profile, by some population biologists and paleoanthropologists have contemplated it as an instance of composite actions of a founder effect, genetic isolation and high extent of inbreeding coefficient (Morwood and van Oosterzee, 2007; Henneberg and Schofield, 2008). In 2008, a team of paleoanthropologists, under the mentorship of Lee R. Berger have excavated the identical size and shape of the human skeletons, and reportedly emerged around the same time of Flores human, from Palau, Micronesia, those small-sized human skeletons from Palau have been recognized as the earliest members of the modern human beings (*Homo sapiens sapiens*), contemplated to be the founder effect, caused by Laron syndrome, occurred in that Palauan subpopulations in that period (Hershkovitz et al, 2007). A group of evolutionists has advocated in favor of genetic drift rather than natural selection as a more effective mechanism in speciation and they cite the reference of "demes" that give rise to the successful emergence of new species. If we would go through a brief review of mode of speciation, we would see that in case of geographic or allopatric speciation, the biological population got completely isolated either due to changes of the climate regime, plate tectonics, sea-level changes or even extreme level of environmental changes, that lead the emergence of that isolated population to form a new species. In the case of peripatric or parapatric speciation, the small segment of a large-sized biological population entered an isolated niche and become genetically isolated, that lead to the formation of a new species. Whereas in the case of sympatric speciation, the emergence of new species or speciation happens from the ancestral population as a result of reproductive isolation within the ancestral population without the occurrence of geographic isolation, and both species cohabit the same geographical location. So, it appears that different modes of speciation, supposedly rely upon genetic drift and other mechanisms of evolution rather than its sole reliance upon natural selection as a driving impetus of biological evolution as it seems to be a long duration process in respect to temporal scale, that perceived to take millions of years to emerge a new species and the comparative studies of the molecular clock of hominoids and hominins, did not fit in that wide span of molecular time clock and the emergence of hominins splitting out of the ancestral lineage of hominoids were perceived to happen in Pliocene (around 4 million years ago, when modern hominins split out of its phylogenetic cousin, chimpanzees). The composite researches, made by molecular geneticists, genomicists, and paleoanthropologists have rendered the act of genetic drift along with the substantial influence of niche construction, genetic introgression, selective breeding, etc., played the key mechanisms in changing the genome profile of the modern human species to narrate the evolutionary history to the present level. And based on these newly emerged hypotheses, evidence, and interpretations, regarding the new narrative, scientists have firmly stated:

"All of these explanations of evolutionary processes contrasting with plain natural selection are at significant variance with the currently still dominant paleoanthropological and archaeological model of recent human evolution, the replacement hypothesis ('African Eve') and the various other Out-of-Africa (OoA) models, which have governed most relevant discourse over the past few decades" (Bednarik, 2011). On the other hand, several scientists have raised concerns whether in absence of substantial paleoanthropological, archaeological and genetic evidence, the sheer consideration of "African Eve" hypothesis or "Out-of-Africa" hypothesis (though both hypotheses have logically presented its narrative in elucidating the emergence of modern human species uniquely, seems to be a credible explanation to a large part of the community of science for few last decades), which is mainly supported by the act of mitochondrial mutation, in elucidating the evolution of modern human species to be justified or not and if it is justified, up to what extent? Naturally, some evolutionists have expressed their concern loudly as they have stated: "This does not mean that there was a single female from whom we are all descended, but rather that out of a population numbering perhaps several thousand, by chance, only one set of mitochondrial genes was passed on. (This finding, perhaps the most surprising to us, is the least disputed by population geneticists and others familiar with genetic drift and other manifestations of the laws of probability.)" (Curtis and Barnes 1989).

A large number of evolutionists, paleoanthropologists, molecular geneticists and genome biologists have contemplated that it's not the sole act of natural selection, one of the sustainable but slow mechanisms of genetic drifting, but the short-time consuming random mechanism of biological evolution rendered expediting speciation of modern human species on Earth (Bednarik, 2011). According to a large number of evolutionists, a sub-population in Sub-Saharan region must have been genetically isolated from the genome of the rest of the human population on Earth, leading to a rapid genetic change of that subpopulation that lead to an evolution of a new species, which does not seem to be inter-fertile with other Africans, and this is construed to be one of the feasible explanations of the emergence of the modern human species on Earth. Naturally, it tried to promote the hypothetical idea of a bottleneck to uphold a biased notion of the historical existence of a primordial woman (Eve), all humans in Holocene are directly or indirectly related to this historic woman (Bednarik, 2011). The critics of the replacement hypothesis have criticized this hypothesis and the supporters of this hypothesis by arguing further that most likely the supporters of replacement hypothesis have perceived and promoted the occurrence of a genetic bottleneck to establish Eve as the mother of humanity, which make the concerned hypothesis weak as the population biologists have noticed that genetic bottlenecks have a substantial negative impact for speciation for a particular species, so the emergence of that species as a leading species among all extant species on Earth seems to be most unlikely as the occurrence of genetic bottleneck brought down the fitness of any biological population (Hawks et al, 2000; Berger, 2008). The critics of the replacement hypothesis pointed out that without infusing the idea of a genetic bottleneck, this hypothesis itself insupportable but with the inclusion of genetic bottleneck option, it did not create any room to accommodate the principles of genetic drift, and alternately forming the genetic trees (Bednarik, 2011). The evolutionists have contended that "genetic trees," those being created without any scientific justification, promoted the false assumption about the exact rate of genetic changes of the concerned population and the improper time of divergence of the concerned population, which split out (time of divergence) of the ancestral lineage. The molecular geneticists and phyllogeneticists have noticed that the time of divergence, indicated by the age-dating molecular clocking technique, examined by the nuclear DNA, mitochondrial DNA and the DNA sample of non-recombining type, extracted from the Y chromosome varies from one another and the difference among three results have found to be so vast it's difficult to infer which result is the appropriate one (Hammer, 1995). The mitochondrial Eve hypothesis has also been criticized in the same way as the critics of this hypothesis have contended that Cann et al (1987), who has tried to promote "exclusive transference of mitochondria through maternal lineage" has no scientific basis, likewise, the supporter of this hypothesis also assumed that the "constancy of mutation rate affecting mtDNA" (Rodriguez-Trelles et al, 2001). The geneticists have observed furthermore that contamination of mtDNA has often been mixed up with paternal DNA, happening most often in case of extant species, altered the resulting index of divergence of the concerned species (Morris and Lightowlers, 2000). Several scientists have noticed that determining the age of divergence of any species by molecular dating protocol, through using the mitochondrial genetic clock has often yielded a bias or improper result by projecting an overestimation and that rendered the radical evolutionists to denounce "mitochondrial molecular clock" as a pseudo-scientific protocol in substantiating the divergence of any derived species out of the ancestral lineage, and determining its tentative time of divergence as well (Rodriguez-Trelles et al, 2002).

We cannot judge any situation by considering the facts and figures, presented in favor or simply presented against such a hypothetical notion. To accept or discard any hypothetical proposition or model, there should be a balanced review in its favor and against it to accept it or reject it further. The geneticists, the YEC protagonists, have simulated historical evolutionary populations (as per biblical narrative) where the creation of Adam and Eve happened miraculously as a maiden human couple, emerging as heterozygous by design, gone through a single-generation of bottle-neck (Noah's Flood, as per biblical narrative) and in course of completion of 200 generation cycles, the modern human species supposedly emerged as a result of the exponential surge of the population after a bottle-neck incident (Sanford et al, 2018). In order to demonstrate experimentally, the scientists have started the simulation run with two individuals, then simulating rapid growth phase till the 10th generation, followed by a bottleneck to 6 people, then running the next phase of simulation to end it to the surge of the population of 1000 people level in the next 200 generations and the scientists observed that a substantial genetic drift was able to generate the cardinal level of allele distribution. For determining and comparing the allele frequency, the published data of the 1000 Genome project for chromosome 22 (which is autosome), chromosome Y and Chromosome M (Mitochondrial chromosome) had been taken into consideration by these geneticists (1000 Genome project,2015; Sanford et al, 2018). At the end of running the simulation, the scientists had not felt that the distribution of alleles appeared to be compatible to the observed allele distribution of chromosome 22, though the geneticists have ascertained that the simulation results have strongly complied the observed allele distributions of chromosome Y and chromosome M (Sanford et al, 2018). In support of their observation, the YEC geneticists have defined that to initiate the early level of heterozygosity in the primordial couple in Eden, the existence of 4 copies of autosomes supposedly existed, whereas they would have existed on progenitor Y chromosome and one progenitor M chromosome, that presumptive existence of autosomes supposedly responsible for designed heterozygosity, whereas chromosome Y and chromosome M would have been considered to be non-variant. The critical review of the results of the simulation run, made on chromosome 22, chromosome Y and chromosome M reveal that the four pairs of erstwhile Edenic chromosomes were not only lack of designed variants they were lack of enough SNP (Single Nucleotide Polymorphisms) alleles and such asymmetrical distributions were triggered by the low level of allele frequencies (Hart et al, 2004; Buggs, April 14, 2018; Sanford et al, 2018). Furthermore, the scientists opined that such level of allele frequencies was evidenced among the young and non-variant chromosomes, those were young from an evolutionary perspective and it had extrapolated from it that the autosomes (like chromosome 22 or autosome 22) having the lower extent of designed variants, were considered to be evolutionary young. The most interesting part of these simulation experiments was shared by the scientists that allele distribution of Adam and Eve, were found to be identical to the allele distributions of modern human populations, as per evolutionary models (1000 Genome Projects, 2015; Sanford et al, 2018). In this part, the observation and comparative analysis of the YEC geneticists on the allele distributions of evolutionary and biblical models are quite intriguing to elucidate the role of a genetic bottleneck, that put both models closer to each other in debriefing the feasible narration of emergence of humanity out of the cradle of primordial, and historical couple in Eden as they stated: "*Both evolutionists and creationists can invoke hypothetical mechanisms to bring their simulated curves into closer alignment with the actual data. ...To reconcile evolutionary simulations with the real data, evolutionists need to invoke a long-term population bottleneck in the distant past....Similarly, biblical simulations require at least one severe population size constriction While, both the evolutionary and biblical models require bottlenecking, for the evolutionary model to invoke a bottleneck is entirely post hoc, while for the creation model bottlenecking is inherent and integral.*" (Sanford et al, 2018).

The acceptance or rejection of any scientific model depends upon the logical analysis and interpretation of it rather than analysis of the direct evidence, as non-availability of that have often stumbled bottleneck of direct investigation. So, whenever the evolutionists have upheld their views that the emergence of modern humanity (which has a wide diversity and exponential growth of population) out of two primordial couple in 200 generations seems to be an absurd idea as the emergence of allele frequency distributions found to be triggered by the combined action of random mutations and genetic drift, that normally require a long period. So, the YEC scientists have considered it and looked for a logical solution that there must be the process that has not been addressed in their interpretation, that would supposedly cut the long span short to generate a high level of allele frequency in a short period and they contend that generations of optimum level of SNPs, would shorten the long period short (though SNPs have been considered as created alleles, rather than mutational levels) (Hart et al, 2004; Jeanson and Tomkens, April 1, 2018; Sanford et al, 2018).

According to the population biologists, the occurrence of genetic drift happens intermittently generations after generations as a result of fractional sampling fluctuations in the gamete gene pool. The YEC geneticists have come up with their observations that for the higher group of organisms, the global population is likely being integrated to substantial demographic forces, more potential than genetic drift that supposedly triggers the emergence of allele frequency at a faster rate (Hart et al, 2004; Sanford et al, 2018). The scientists observed furthermore that the main reason for a quick alteration of allele patterns has chiefly triggered by the demographic factors, rather than induced by the randomized occurrence of genetic drifts. The scientists have referred that the demographic population equilibrium has been in dynamic mode due to several factors like war (life first world war), historical conquests (like an exploration of new World by Vasco Da Gama and establishing new settlement of Europeans in North America), spreading out of disease (like the occurrence of pandemics like plagues, yellow fever), solo-religious practices in some geographical locations, render the population explosion and that not only enhanced the level of genetic changes in the population gene pool, trigger the extent of genetic drift as well as accelerate the changes of allele frequency too. In a recently published paper on the simulation investigations, meticulously studied by a group of 92 scientists observed that a "rapid and massive" level of global demographic shifts happened in the historic human population, and such demographic shifts would have created a short–span, widespread spreading of an allele in the global population of the modern human race (Narasimhan et al, 2018).

The YEC geneticists have experimentally demonstrated that genetic drift has a higher level of functional potential that depends upon the change of demographic factors in a population and to present a simple analogical elucidation, the scientists referred to an example of diffusion event of a colored dye in water, where the colored dye usually diffuse in the water very slowly but any external stimulation like stirring the water would accelerate the diffusion process and the dye homogeneously diffused in the water much faster, likewise, any stirring in the gene pool of the global population, which is recognized as "demographic stirring" would accelerate the functional potential of genetic drift, that helps to run flawless genetic simulations, and undertake a better population modeling to narrate the emergence of any derived group of an organism out of the cradle of an ancestor in a feasible manner. The scientists have observed further that to undertake the important factor like demographic stirring in accelerating the functional potential of genetic drift and to run a flawless genetic simulation by adopting a better population model, it requires to count this correction factor. The YEC geneticists have mentioned that instead of conventional assumptions of the erstwhile human population size, that was considered to be comprised of 10,000 individuals (that supposedly gave rise to the modern human race nowadays), they have simulated much smaller size of the population of 1000 individuals and the important demographic factor "demographic stirring" is missing in their simulation model of the human population, propounded by the YEC geneticists, that hinders the acceleration of allele spreading among the human population. So there is a simple analogy needs to be taken into consideration, where pure gold (which is too soft for ornaments) is unsuitable to use without mixing alloys to a certain extent to make it harder like the classical occurrence of genetic drift in a gene pool of human population, without demographic stirring it remained to be an ineffective factor to change the alleles frequencies, those were considered to be natural selection, across the global population (Sanford et al, 2018).

The YEC geneticists have approved the shortcomings in their numerical simulations, where large size of the population is a big impediment in running the model population as they affirmed (Sanford et al, 2018): "*In particular, we cannot model the billions of very rare alleles that are now accumulating in our very large population...Our simulations fail to model the vast number of rare alleles that would accelerate as the human population grew rapidly from thousands of people to billions of people. Realistically, any type of demographic stirring would draw large numbers of SNPs into the 1-99% frequency range, helping to explain a large number of actually observed human polymorphisms. Therefore, there is an enormous reservoir of SNPs in the actual human population that Mendel is neither currently simulating nor plotting.*" However, the YEC geneticists have briefly elucidated their findings in three segments, after successfully running the simulation, that rendered them to reconcile the Biblical Adam and Eve supposedly possessing human allele distribution and the main observation of these scientists are as follows (Gauger, 2012; Sanford et al, 2018):

A. Adam and Eve had been emerged as a genetically heterozygous historic couple in Eden, subjected to the population constriction after few generations of their emergence, that demographic stir enhanced the extent of genetic drift: The YEC geneticists have observed that their simulations of heterozygous Adam and Eve seem to be a simplified model so, based on the preliminary simulations, the designed diversity model appeared to be discordant to the distribution of observed allele frequency. In this model several factors that influence the allele frequency distribution are as follows:

i. The duration and number of bottlenecks, the ancestral population has gone through to emerge in derived form.

ii. The ratio of designed alleles and mutational alleles in the population and

iii. The ratio of designed alleles in distinct allele frequency classes. The careful consideration of these variables helped the scientists to simulate to yield allele frequency distributions that is found to be identical to the observed biological allele frequency distributions of modern human populations.

The YEC geneticists have further contended that they found that in their simulation run, it requires fine-tuning of several variables to optimize the index of allele frequency and the simulation result would comply the biblical narrative of the emergence of modern humanity, supposedly initiated out of the primordial stock of Adam and Eve by –

a. By coordinating fine adjustments of the ratio of mutational alleles versus designed alleles.

b. By simulating a growth pause after the biblical chaotic event of Babel. And

c. Simulating an event of the accelerated rate of mutation accumulation in the 25-50 generations in the post-flood regime.

The YEC geneticists have been critically observed that in their numerical simulation the yield of allele distribution has emerged with a range of gap of 3-20% range, that stunted morphing the graphical plotting of allele frequency distribution by truncating the graphical line of curvature and the accelerated rate of genetic drift (the demographic stirring) would fill-up theses gaps to present a picture of even spreading out of allele frequency in the concerned population model. In a contemporary experimental observation on population genetics, the geneticists have meticulously observed that the event of accelerated genetic drift happens in a small size of the population and the act of "classical genetic drift" had found to be stalling the spreading out the allele frequency distribution, when the size of the population surged the level of 1000 individuals or more (Carter and Powell, 2016). So, empirical investigations have indicated that for an even spreading out of the allele frequency distribution, the small size of the population, that remained smaller for several generations is pre-requisite. Eventually, this is the strong supporting point for the YEC geneticists who have come up with the biblical narrative of the beginning of humanity out of two primordial couple of Adam and Eve in the Eden, followed by the post-flood bottleneck event, where population sized restricted to six and subsequently followed by a restriction of population growth of the tribes following the chaotic or violent dispersion out of Babel (Carter and Hardy, 2015; Sanford et al, 2018).

The mix and match variables of mutational and designed alleles, in their experimental trial of simulating to secure an even spreading of allele frequency by YEC geneticists, found to be quite interesting as the combination of mutational alleles with designed alleles (starting with an allele frequency of 25%), rendered the drifting of designed alleles towards low allele frequencies and found to fill up the 3-20% gap, that stalled the graphical presentation of even spreading out the allele frequencies in the population. Whereas a composite input of two or more types of designed alleles (either starting with an allele frequency of 25% or 50%) has helped in simulating a better way (Gauger, 2012; Sanford et al, 2018).

B. Adam and Eve had been emerged with inherent designed diversity, along with a range of demographic forces rather than usual occurrence of classical genetic drift: The YEC geneticists have perceived the idea of inclusion of important factors like selective sweep, that accelerates the frequency of beneficial mutations, immediately followed by fixation of it leading the possibility of reduction of genetic variation in the DNA sequences, close to mutation site and differential expansion of sub-populations to render the optimization of simulation biologically. If we look at the functional roleplay of natural selection, one of the important evolutionary mechanisms, we realize that "natural selection" itself selectively chose the favorable alleles for any environment to make sure that a set of consistent traits have been passed on for a species from parents to offsprings from generation to generation intermittently. So, through the loop of natural selection, if any species possess more favorable alleles, it seems to be capable to sustain for a long period and capable of passing on the genetically consistent traits from ancestors to the descendants of any sexually reproduced species that form a viable population size. In the field of evolutionary genetics, "selective sweep" or genetic hitchhiking is a popular terminology that elucidates how main and associated alleles (located on the same location of a chromosome), those recognized to trigger favorable adaptations and frequency distribution of which has been witnessed in the population (Kim, 2006). So, the occurrence of genetic drift in the gene pool of any biological population has a close analogy to a time-consuming diffusion process in a marine biological perspective, which has been recognized to be a passive process that would not change the allele frequencies of any population, unless the size of the population was very small. The YEC geneticists have considered that during their simulation run through Mendel 2.7.2, they have assumed that the alleles are neutral, so those have not been screened under the strainer of natural selection, and it has further been perceived that emergence of global populations happened as a stochastic mating of individuals among historic populations. So several alternative demographic factors have appeared to drive up the acceleration of allele frequency distributions across the population in a short period as well as it expedited the formation or surging of a genetically diverse population in the small phase of time, and those factors include an exponential surge of population in post-bottleneck regimes, selective sweeps, fragmentation of large populations to sub-populations, migration or invasion related changes of population dynamics, extinction of genetically fragile individuals of scattered sub-populations, that don't fit into a climatic regime of any given environment, etc. (Coop and Ralph, 2012). By promoting the idea of selective sweeps, the YEC geneticists have tried to give a feasible alternative of genetic drift that expedite the spreading of allele frequencies across the population and the extrapolation of it indicates further that a sequential occurrence of selective sweeps would easily fill in "the gap" of the distribution of allele frequency spreading of it. It rendered the YEC geneticists to contemplate that their conceptual narration of the designed diversity of the primordial population has complied with the principle of the rapid mode of adaptation to local climatic regimes of the micro-environment (Sanford et al, 2018). The YEC geneticists have tried to elucidate the biblical notion of "fill the Earth" has seemed to have resembled the evolutionary narrative of "fill all the environmental niches in the Earth" as a result of "rapid adaptive population fragmentation" (Coop and Ralph, 2012; Sanford et al, 2018). The scientists have contemplated further that possession of designed alleles in any biological species found to effectively render it to fit in the selection-driven adaptation as the availability of the genetic variants are found to be there from first generation (F1) so the chance of wasting time in the evolution of mutational allele followed by its movement from "allelic near-extinction" to "allelic fixation" through in course of mutation is less likely (Coop and Ralph, 2012). The scientists dealing with this simulation run contended further that (Caballero and Garcia-Dorado, 2013; Greenbaum et al, 2014; Sanford et al, 2018)-

i. Existence of such designed alleles in high-frequency level have been witnessed in the form of functional linkage in the ancestral lineage of human species make it fit for rapid adaptation.
ii. When a pair of "high impact allele" is concerned, the favorable allele (the major or leading one) is found to be more befitting or more adaptive to the particular habitat, the species lived in respect to the unfavorable or minor alleles. So, the chance of rapid increase of the number and even spreading of the frequency of the favorable allele in the population and its fixation in the next level is more likely, whereas the chance of minor or unfavorable allele to be zeroed out in the population gene pool is there.
iii. However, the high impact alleles have formed bondage to the "almost-neutral variants" during such even spreading out mechanisms of acceleration of allele frequency distribution across the population that kind of genetic maneuvering has technically recognized as "selective sweeping." So, in the modified simulation run (200 generations) for a short duration, of the flawless allele frequency distribution with small population model, the scientists (YEC geneticists) have noticed that in their graphical presentation the allele distribution has appeared as a smooth portrayal of allele distribution without any visible gap and the allele variants of "selective sweep" be triggered by the combination of designed alleles to a greater extent along-with mutational alleles to a lesser extent.

Genetically, mutations are generally found to be deleterious or found to be the less beneficial process, in mutational alleles selective sweeps would have been expected to be a rare occurrence in comparison to designed alleles, selective sweeps seem to be very common and designed alleles would be inadequately present in the first generation of the ancestral lineage of modern human species. The YEC geneticists have observed that when the major and minor allele variants of a mutational allele have appeared to be functionally antagonistic, the allele variants of designed alleles are found to be complementary functional roles as one favored over the other (Sanford et al, 2018; Jeanson, 2019). The designed alleles are a supposedly forming pair with nearly-neutral alleles and existed in form of linkage blocks that likely to the emergence of the species, evolved with inherent polygenic traits. The YEC geneticists have also agreed that they don't want to give a false empirical projection that all human alleles are found to be in an existing state as large frequencies of alleles have supposedly been diminished since the journey of humanity started since its historic evolution as the extant descendants of the ancestral predecessors should be considered to be the "lucky lineages," carry forward the allele frequencies of the ancestors, leaving behind the lineages of extinct populations, the diversified alleles, those have been possessed by them, supposedly disappeared after being lost in the verge of extinction (Helgason et al, 2003). At the same time, the exponential surge of the population had enhanced the emergence of some rare mutational alleles that helped to offset the depletion of an ancestral treasure trove of alleles.

The primary objective of the YEC geneticists of carrying out of such numerical simulation is to tally the simulation results with observed allele distributions of chromosome number 22 or autosome number 22, yet the scientists have reviewed the allele distributions of chromosome Y and mitochondrial chromosome of modern human species (Carter et al, 2018; Sanford et al, 2018). The scientists have observed that though Chromosome Y and Chromosome M are found to be distinct in structure and function, their distributions of allele frequencies are found to be identical and both of these have been characterized by identical traits like "very few high-frequency polymorphisms." From the genetic perspective, the polymorphic alleles have emerged with very low-frequency level and the scientists have ascertained that such features have compliance with the heterozygous Adam and Eve model in Eden (Carter et al, 2018; Sanford et al, 2018). Based on their experimental simulation, the YEC geneticists have come up with a radical explanation about the number of copies ancestral chromosomes, that existed in Eden, possessed by Adam and Eve as scientists claimed that in Eden there was the existence of two special chromosomes as single ancestral copies, instead of four copies and the scientists have further claimed that the ancestral Edenic chromosomes, have had "zero designed alleles" and the observed polymorphisms would have possibly yielded in course of accumulation of mutations, rather than triggered by designed alleles. The YEC geneticists have firmly claimed that if and when the modern human genome would have stripped off designed alleles, the allele distributions of human chromosomes would be supposedly identical to the allele distributional pattern to chromosome Y and chromosome M (Carter et al, 2018; Sanford et al, 2018). As the allele distribution patterns of chromosome Y and chromosome M have projected it as young chromosomes, it rendered the YEC geneticists to assume that in the young state these two chromosomes have had too little time the accumulation of function of several mutational alleles that rendered it to drift to higher frequencies (Sanford et al, 2018).

C. Adam and Eve's gametogonia possessed designed allele diversity: The YEC geneticists have contemplated historic creation of a vast number of gametogonia[7] in Adam and Eve in Eden. It has been perceived by the YEC geneticists that each gamete in Eden has been designed with its unique set of alleles and unique linkage types the original population of reproductive cells, representing a gene pool, could have been resembled a large-sized human population. The scientists have simulated to elucidate how the designed gametes might have gave rise to modern allele frequencies. To run the designed gametes simulations model, the scientists chose a small population size, comprised of 50 offsprings of Adam and Eve, and these 50 offsprings were perceived to be derived from 100 designed gametes (gametogonia) of Adam and Eve (Carter et al, 2018; Sanford et al, 2018). The scientists have affirmed further that these biblical population of 50 individuals, as well as offsprings of Adam and Eve, had been designed with 989,000 or more allele pairs of designed alleles. In the first phase of the simulation run (through Mendel Go version), the scientists allowed these initial population to grow till 9th generation, followed by a bottle-neck (bottleneck of genesis Flood), that shrunk the size of the population to 3 pair of reproducing couples (or 6 individuals. Comprised of 3 adult men and 3 adult women), followed by a surge of a population to reach at the level of 1000 individuals and the run of

---

[7] Gametogonia: These are stem cells, give rise to formation of gametes or sexually differentiated germ cells, possessed in gonads (the reproductive organs that produce gametes and hormones). The male gametogonia or spermatogonia (that produce sperms) have found to be located in testes, whereas the female gametogonia of oogonia (that produce eggs) have found to be located in ovary

the simulation stopped at 200th generation (Carter et al, 2018; Sanford et al, 2018). The scientists have found that at the end of this simulation the composite allele frequencies (that ranges from 1-50% level) showed (comprised of mutation alleles, favorable designed alleles, and unfavorable designed alleles) having nearly neutral in fitness effect and the experimental results of designed gametes allele distribution found to be identical to the observed distribution (Sanford et al, 2018). According to the "Designed Gametes Model," the functional allele variants itself would not have such potential to create a wide gene pool of allelic diversity that might create a genetically diverse large population and ensure its sustainability either so these functional variants have been contemplated to form the "integrated linkage blocks," that supposedly ensures the genetic diversity of the contemporary surged population of the modern human species on Earth. However, at the end of their studies, the YEC geneticists have firmly perceived that their research elaborates that if Adam and Eve in Eden would have been endowed with the optimum level of "designed genetic diversity," it might give rise to the contemporary level of allele frequency, observed among the global population of modern human species (Thistletwaite, 2003; Sanford et al, 2018). According to the YEC geneticists, the initial part of this journey of the history of humanity on Earth has must be considered to be "pre-program designed diversity into the genomes of the first couple" by the supernatural body, which has considered to be the glory of God (according to the biblical narrative) (Thistletwaite, 2003; Sanford et al, 2018). It has further been elucidated by the YEC geneticists that if the journey of humanity had been started with an act of random mutation as well as triggered by mutational alleles, it might take a long time as the act of random mutation is by and large deleterious, so it's difficult to get in touch of beneficial type of variants followed by fixation of it, that is hypothetically long time taking event, and it is found to be almost impossible to witness the exponential surge of genetically diverse, modern human species on Earth nowadays, that has conveniently short span of evolutionary time of divergence, in respect to other vertebrates and mammals, unless the historical progenitor of it had not been endowed with designed allele diversity in its genome, the biological holy grail. The first part of the history of humanity seems to be a stochastic process, as it entangled with the reconciliation of biblical narrative of the creation of humanity and transcending genetic inheritance from ancestral lineage to the derived form of the modern human race and debriefing the genetical interpretation of inheritance and distribution of allele diversity in the modern human population; whereas the second part of it is found to be fitting in heuristic format, where cause and effect of the evolutionary journey of modern human have been narrated conveniently, under the narratives of evolutionary biology. The YEC geneticists have firmly affirmed their observation and interpretation after carrying out successful completion of their simulation studies on this designed gametes model experiment as they have stated "*The designed gametes model appears to be especially robust, and in our opinion is even elegant. It seems to be the best explanation for how Adam and Eve might have simultaneously given rise to our current human allele patterns and our current linkage patterns*" (Sanford et al, 2018). Furthermore, the YEC geneticists have narrated that after coming across the enigmatic, initial part of the journey of humanity, the even spreading out of allele frequency, accomplishment of which has successfully been witnessed in the global population of the modern human population, found to be assisted by a range of demographic and genetic factors, ranging from accelerated genetic drift, selective sweeps, change of population dynamics due to migration and invasion, occurrences of pandemics, sub-population surges, lineage extinctions, etc. (Thistletwaite, 2003; Sanford et al, 2018).

References

Bednarik, R.G. (2011). Genetic drift in recent human evolution? In: Urbano, K.V. (ed). *Advances in Genetic Research.* Vol 6. Pp. 109-160.

Berger, W. H. (2008). Sea level in the late Quaternary: patterns of variation and implications, *Int J Earth Sci.* 97: 1143-1150.

Buggs, R. (April 14, 2018). Adam and Eve: lessons learned. http://richardbuggs.com/2018/04/18/adam-and-eve-lessons-learned/

Caballero, A., and García-Dorado, A. (2013). Allelic diversity and its implications for the rate of adaptation. *Genetics.*195(4): 1373-84. doi: 10.1534/genetics.113.158410.

Cann, R.L., Stoneking, M. and Wilson, A.C. (1987). Mitochondrial DNA and human evolution. Nature. 325: 31-36.

Carter, R. and Hardy, C. (2015). Modelling biblical human population growth. *Journal of Creation* 29(1):72–79.

Carter R., and Powell, M. (2016). The genetic effects of the population bottleneck associated with the Genesis Flood. *Journal of Creation* 30(2):102–111.

Carter, R.W., Lee, S.S. and Sanford, J.C. (2018). An overview of the independent histories of the human Y-chromosome and the human mitochondrial chromosome. In *Proceedings of the Eighth International Conference on Creationism*, ed. J.H. Whitmore, pp. 133–151. Pittsburgh, Pennsylvania: Creation Science Fellowship.

Coop, G. and Ralph, P. (2012). Patterns of neutral diversity under general models of selective sweeps. *Genetics* 192(1): 205-224.

Curtis, H., and Barnes, N. S. (1989). *Biology. Fifth edition.* W.H. Freeman. P. 1192.

*Darwin, C. (1859). On the Origin of Species by Means of Natural Selection, or the Preservation of Favoured Races in the Struggle for Life. (*Full image view *1st ed.), London: John Murray, p. 502.*

Gauger, A. (2012). Adam and Eve Redux. *Christian Research Journal* 35(1): 1-6. http://www. equip.org. christian-research-journal/

Greenbaum, G., Templeton, A.R., Zarmi, Y., and Bar-David, S. (2014). Allelic richness following population founding events--a stochastic modeling framework incorporating gene flow and genetic drift. *PLoS One.* 2014 Dec 19;9(12):e115203. doi: 10.1371/journal.pone.0115203.

Hammer, M. (1995). A recent common ancestry for human Y chromosomes. *Nature* 378: 376–378. https://doi.org/10.1038/378376a0

Hart, J.R., Johnson, M.D. and Barton, J.K. (2004). Single nucleotide polymorphism discovery by targeted DNA photocleavage. *PNAS USA* 101(39): 14040-4.

Hawks, J., Hunley, K., Lee, S-H., and Wolpoff, M. (2000). Population Bottlenecks and Pleistocene Human Evolution, *Molecular Biology and Evolution*, 17(1): 2–22,

https://doi.org/10.1093/oxfordjournals. molbev.a026233

Helgason, A., Nicholson, G., Stefansson, K., and Donnelly, P. (2003). A reassessment of genetic diversity in Icelanders: strong evidence from multiple loci for relative homogeneity caused by genetic drift. *Ann Hum Genet* 67 (4): 281–297.

Henneberg, M. and Schofield, J. (2008). *The Hobbit Trap: Money, fame, science and the discovery of a 'new species.'* Wakefield Press. P. 176.

Hershkovitz, I., Kornreich, L., and Laron, Z. (2007). Comparative skeletal features between *Homo floresiensis* and patients with primary growth hormone insensitivity (Laron Syndrome). *Am J Phys Anthropol.* 134(2): 198-208. doi: 10.1002/ajpa.20655.

Jeanson, N.T. (2019). Testing the Predictions of the Young-Earth Y Chromosome Molecular Clock: Population Growth Curves Confirm the Recent Origin of Human Y Chromosome Differences.

https://answersingenesis. org/theory- of-evolution/molecular-clock/testing-predictions-human-y- chromosome-molecular-clock/

Jeanson, N.T. and Tomkens, J.P. (April 1, 2018). Genetics confirms the recent, supernatural creation of Adam and Eve. Featured in "*Searching for Adam.*" https://answersingenesis.org/adam-and-eve/genetics-confirms-recent-supernatural-creation-adam-and-eve/

Kim, Y. (2006). Allele frequency distribution under recurrent selective sweeps. *Genetics* 172: 1967–1978.

Kimura, M. (1968). Evolutionary rate at the molecular level. *Nature* 217: 624–626. doi:10.1038/217624a0

Kimura, M. (1983). *The neutral theory of molecular evolution.* Cambridge University Press. ISBN 978-0-521-31793-1.

Morris, A.A.M. and Lightowlers, R.N. (2000). Can paternal mtDNA be inherited? *Lancet* 355: 1290-1291.

Morwood, M. J., and van Oosterzee, P. (2007). *A new human: the startling discovery and the strange story of the "Hobbits" of Flores, Indonesia.* New York: Harper Collins. P.256.

Narasimhan, V.M., Patterson, N., Moorjani, P., Lazaridis, I., Lipson, M., Mallick, S., Rohland, N., Bernardos, R., Kim, A.M., Nakatsuka, N., Olalde, I., Coppa, A., Mallory, J., Moiseyev, V., Monge, J., Olivieri, L., Adamski, N., Broomandkhoshbacht, N., Candilio, F., Cheronet, O., Culleton, B., Ferry, M., Fernandes, D., Gamarra, B., Gaudio, D., Hajdinjak, M., Harney, É., Harper, T., Keating, D., Lawson, A.M., Michel, M., Novak, M., Oppenheimer, J., Rai, N., Sirak, K., Slon, V., Stewardson, K., Zhang, Z., Akhatov, G., Bagashev, A.N., Baitanayev, B., Bonora, G., Chikisheva, T., Derevianko, A., Dmitry, E., Douka, K., Dubova, N., Epimakhov, A.,

Freilich, S., Fuller, D., Goryachev, A., Gromov, A., Hanks, B., Judd, M., Kazizov, E., Khokhlov, A., Kitov, E., Kupriyanova, E., Kuznetsov, P., Luiselli, D., Maksudov, F., Meiklejohn, C., Merrett, D., Micheli, R., Mochalov, O., Muhammed, Z., Mustafokulov, S., Nayak, A., Petrovna, R.M., Pettener, D., Potts, R., Razhev, D., Sarno, S., Sikhymbaeva, K., Slepchenko, S., Stepanova, N., Svyatko, S., Vasilyev, S., Vidale, M., Voyakin, D., Yermolayeva, A., Zubova, A., Shinde, V., Lalueza-Fox, C., Meyer, M., Anthony, D., Boivin, N., Thangaraj, K., Kennett, D., Frachetti, M., Pinhasi, R., and Reich, D. (2018). The Genomic Formation of South and Central Asia. *bioRxiv*.

Rodriguez-Trelles, F., Tarrio, R., and Ayala, F.J. (2001). Xanthine dehydrogenase (XDH): episodic evolution of a 'neutral' protein. *J. Mol. Evol.* 53(4-5): 485-495.

Rodriguez-Trelles, F., Tarrio, R. and Ayala, F. J. (2002). A methodological bias toward overestimation of molecular evolutionary time scales. *PNAS USA* 99(12): 8112-8115.

Sanford, J., Carter, R., Brewer, W., Baumgardner, J., B. Potter, and Potter, J. (2018). Adam and Eve, designed diversity, and allele frequencies. In *Proceedings of the Eighth International Conference on Creationism*, ed. J.H. Whitmore, pp. 200–216. Pittsburgh, Pennsylvania: Creation Science Fellowship.

Suzuki, D.T., Griffiths, A.J.F., Miller, J.H., and Lewontin, R.C. (1989). *An introduction to genetic analysis*. W.H. Freeman Co. Ltd., 4th Edn.

The 1000 Genomes Project Consortium. (2015) A global reference for human genetic variation. *Nature* 526: 68–74. https://doi.org/10.1038/nature15393

Thistletwaite, S. B. (ed.) (2003). *Adam, Eve and the Genome: The human genome project and theology*. Minneapolis. P. 200

Wright, S. (1929). "The Evolution of Dominance." *The American Naturalist*. 63 (689): 556–561. doi:10.1086/280290

Chapter 9

A Brief Review on Designed Diversity, Adam and Eve, Human Allele Frequency: Concerning Mutational Alleles and Designed Alleles

..Currently observed human allele frequency patterns could not rise from a single couple. The logic here is that, since there were only four sets of chromosomes in Eden, all variants would have had an initial frequency of either 25%, 50%, or 75%. Today, most allelic variants have frequencies in the range of 0-10%. Therefore, it is claimed that observed human diversity disproves a literal Adam and Eve.

---- *John Sanford & Others*[8]

The fierce battle on the historical existence of Adam and Eve as a primordial couple on Earth and the perceptional conjecture of special creation which promotes the emergence of the modern human species has not only been confined as a matter of socio-religious conflict between the evolutionists and creationists, rather this conflict has witnessed to be brewed in between the orthodox believers of special creationists and theologists, strongly supported by the followers of church and religious institutions and several researchers, scientists, evolutionists, and the communities believe in scientific principles of the history of the emergence of humanity since 19th century. The conflict of perceptions regarding the emergence of humanity became a "hot potato" in the 20th century, once the holy and historical existence of Adam and Eve had initiated challenges from different segments of society (Carter, 2015; Van Doodgewaard, 2015). Several members of a Church going community started gaining strength to challenge the historical existence as well as the miraculous creation of Adam, Eve and no existence of Eden and no historical occurrence of the exodus of the primordial couple

[8] *Others*: Rob carter, Wes Brewer, John Baumgardner, Bruce Potter and Jon Potter.

out of Edenic State by simply contending it as a theological and Biblical narratives, which had a moral and religious impact on humanity to shape it to follow certain social norms and practices of society rather than consider it a feasible interpretation of scientific narratives about our origin and evolution on Earth (Carter, 2017; Venema and McKnight, 2017).

Several theistic evolutionists have started advancing with a series of genetic arguments to prove that the existence of literal Adam and Eve has been contemplated as an allegorical narrative of theology, that has been spun with mythological ideas rather than scientific evidence. It has tacitly been assumed that Adam and Eve would have been created without inherent genetic diversity. And the most argumentative attacks against Biblical genesis of Adam and Eve, as it hardly validates the possibility of the emergence of genetically diversified modern humans out of this primordial couple and this observation and explanations of the theistic evolutionists have been communicated in different public forums and electronic media (Venema, 2010; Venema and McKnight, 2017). So theoretically it has been assumed further that the genetic diversity, which has been observed among the modern human species supposedly accrued in course of random mutations. But such an assumption has found to be pseudoscientific, as an accumulation of such extreme level of mutations in a very short period would have pushed the concerned species on the brink of extinction. So, hypothetical arguments of theistic evolutionists that promote the ideas of a small population of several thousand individuals (serving as a pool of human allele frequency) as the ancestral lineage of modern human species supposed to be more tenable than the theological claim of literal Adam and Eve as the ancestral lineage of the modern human race (Shaffner, 2017a; 2017b).

Several opposing groups of evolutionists have contended further that the entire data analysis of allele frequency should not be considered as the sole criterion to contend or discard in judging the historical models (Myers et al, 2008; Harpak et al, 2016; Baharian and Gravel, 2018). Some scientists have come up with a unique hypothetical model, called "Designed diversity model," where the scientists, specifically the supporters of "Intellectual designs" have contended that emergence of Adam and Eve have supposedly taken place in the heterozygous state and that state has supposedly persisted for more than a decade (Woodmorappe,1996; Wood and Francis, 2016; Jeansons and Tomkins, 2017; Sanford et al, 2018). However, some experimental reviews on "Designed genetic diversity" model by Hossjer et al (2016A, 2016B); Gauger (2018) and the meticulous review of "Initial Contrasting Alleles" or ICA that rendered the scientists to design a numerical simulation program, called "*Mendel's Accountant*" or "*Mendel*" has been scientifically adopted for further review in 2005 onwards by Sanford et al (2018) have emerged out to invalidate erstwhile observation that claimed that the emergence of Adam and Eve would have risen without substantial genetic diversity, rather it has indicated that the maiden human couple ,would have been designed with millions of variable genetic sites. As an experimental model, "*Mendel*" has primarily defined as an accounting program, dealing with an account of numerous genetic transactions to determine genetic loss or gain in its virtual mode of book-keeping that rendered the scientists to give an idea of the hereditary as well as the evolutionary impact of such genetic transactions altering the dynamics of the concerned species or any biological entity, belonging to a certain taxonomic rank. Apparently, "Mendel" has been considered as a typical numerical simulation model, technically ridiculed as "Black Box," as several scientists have expressed their biased views that some accounting results have been found to yield inferior results, but that does not mean the numerical simulation protocol itself is not credible. The credibility of the result of data simulation depends on the efficiency of the model like the feasibility of the result on the final account depends on the sincerity and efficiency of the accountant. So, as a functional model, the result of the genetic accounting of "*Mendel*" should not be underestimated as the scientists and experts formed it (Sanford et al, 2007a, Sanford, et al, 2007b; Sanford and Nelson, 2012; Baumgardner et al, 2013; Rupe and Sanford 2013; Sanford et al, 2015) intended to make a common platform between the creationists and the members, belonging to the community of science to discuss a common issue like historical emergence of the modern human race in the background of genetic variation out of mutational variations or mutational alleles and which has been defined under the new line of thought, elucidated under "Designed Diversity Model" (Sanford et al, 2018). The conventional genetical and evolutionary perceptions contend the idea that the key source of genetic variation comes through the mutations, driven by potential formation and function of mutational alleles, whereas in the "Designed diversity Model," the special creation has been contemplated to be triggered by the impetus of genetic variance, which has been recognized as "designed alleles," rather than conventional effectuation of mutational alleles. So, concept-wise there must be a level of difference between functionalities of mutational alleles, and designed alleles as the accumulation of mutational alleles seem to be time taking and needs to go through a series of generations whereas the designed alleles appeared from like a mechanical copycat descended from generations to generations, unaltered (Sanford et al, 2018). If we try to dig deeper, we could notice that emergence of mutation alleles happen as a result of a typographical error in the genome (imperfect order of base sequence during DNA transcription, that rendered formation, and production of the wrong sequence of amino acids or proteins) and appeared as a single isolated copy, so it should broadly be categorized as a not so beneficial process, whereas the designed alleles remained unaltered, descended vertically through generations found to be consistent and seems to be functionally beneficial than mutational alleles (Lacy, 1987; Sanford et al, 2018). Following the principles of "Designed Diversity Model," when the existence of plenty designed alleles helped in the emergence of a distinct form of a biological entity, mutational alleles are considered to be rare type, that are unlikely to emerge sustainable genetic variations and leading promote an ambiance for speciation in short-run, down the trail of evolution. The scientists observe that with the *n* being a population's size, only one copy of newly emerged mutation allele could get infused in a population of 2n, so there is little chance to have its sustainability as the chance of its self depletion is there due to intermittent effect of genetic drifting, happen in few generations, during its passage from one generation to next (Rupe and Sanford, 2013).

The identity of designed alleles is found to be enigmatic, which needs to be elucidated further. In terms of genetics, specifically in population genetics, the minimal level of genetic variation in any organism has often been triggered by a single letter alteration in its genome, scientifically recognized as "*single nucleotide variants*" or SNVs. If and when the scientists identified minor allele or variant allele (frequency of it is >1% level), it has specifically designated as "Single Nucleotide Polymorphism" or SNPs. In the publication, entitled "*A global reference for human genetic variation,*" which has been conducted and published by "1000 Genome Project Consortium" in 2015, all Single nucleotide variants have been recognized as SNPs, irrespective of their allelic frequency (1000 Genome project Consortium, 2015). The findings of 1000 Genome project is unique as it revealed that there are around 84 million SNPs in the modern human population, 64% of which are rare alleles, having an allelic frequency found to be <5% and which is likely to be an under-estimated figure as the population geneticists have contemplated that each nucleotide of modern human genome supposedly go through innumerable times of mutation in each generation so the expected empirical result of SNPs in human population supposed to be around 3 billion instead (Sanford et al, 2018). Based on the presence of rare alleles in "single people group or sub-populations" in the modern human species, the population biologists considered its emergence through mutation cascades in the recent past (Sanford et al, 2018). The defendants of Intellectual Designs have also construed that apart from the emergence of rare alleles, the current state of genetic diversity supposedly emerged out of designed genetic variants, integrated to genetic variations of Adam and Eve, supposedly giving rise to the modern human race, according to Biblical narrative (Sanford and Carter, 2014). According to the contemporary analysis of 1000 Genomes Project (2015), there are 8 million SNPs, having allelic frequency >5%, most of these alleles have been considered as designed alleles.

The population biologists observed that a single human on average carries 4-5 million SNPs and that represents possession of 30% genetic variation (a lion's share of the genetic variation of the human population) (Levy et al, 2007; Carter, 2018). A considerably high rate of polymorphism has been witnessed among the human beings of African origin (1000 Genomes Group, 2015). The scientists have contended further the SNPs possessed by a single human, found to be an identical number of SNPs, possessed by human populations and it helped them to contemplate that as an individual, each human supposedly possesses almost identical genetic variants, represented by globally dispersed human population (Carter, 2018). So, it helped the population biologists to extrapolate that Adam and Eve as a primordial couple should have contained an identical level of genetic diversity and they supposed to carry most of the 8 million common SNPs, as a modern human couple carries it in its genome nowadays (Sanford and Carter, 2014). The supporters of the Young Earth Creation and Intellectual design hypotheses, who would like to define and redefine the emergence of the human race in the Biblical narrative of Genesis, have contended that the genetic diversity supposed to be lost as a result of the genetic bottleneck of the biblical flood but the population geneticists have ascertained that the high extent of the bottleneck of the smallest population size (population size of two) could be recovered and able to roll back in the pre-bottleneck level of diversity, if the span of bottleneck lasts for few generations followed by the exponential growth of population is there (Nei et al, 1975). So, the geneticists supported YEC hypothesis, have contended that there would have a chance of loss of genetic diversity must be there during the biblical flood but the possibility of roll-back of the pre-flood genetic diverse state is there as well if we consider the fact of the exponential growth of modern human race in a short period. The population biologists have run a computer simulation with a single generation flood-type of a bottleneck with three founding couples and they observed that chance of recovering the lost genetic diversity in the pre-flood state is there with the exponential growth of population in the post-flood era (Carter and Powell, 2016). In two separate occasions, the YEC advocating scientists had come up with two unique findings as one research group revealed that if/ when Noah's three daughters-in-law were distantly related, the simulation results show that the population to refuge in Noah's Ark during the Genesis flood, would have contained 80% of the pre-flood genetic diversity (Carter and Powell, 2016). In another simulation, YEC researcher had presented a unique result where it assumed that if there were a remote possibility where the three sons of Noah, Shem, Ham, and Japeth got married to their sisters (instead of three distantly related brides), the exponential growth of modern human race indicated that the chance of recovery of genetic diversity pre-flood state seems to be around 60% (Carter, 2018). So, the numerical simulation indicated that the biblical flood might have rendered the partial loss of created diversity but Noah's community in Ark supposedly carry millions of polymorphic alleles that had promoted the contemporary emergence of well-diversified modern human species on Earth.

According to the findings of 1000 Genomics Project, (2015), The possession of 8 million alleles, which has been considered as designed alleles and 64 million of SNPs, considered to be mutational alleles in the human genome rendered the modern and genetically diverse human species to sustain on Earth even in any critical bottleneck and this possession of allelic treasure-trove in the human genome, rendered modern human species to be surged as the ultimate mammal with exponential growth to occupy all diversified ecological niches on Earth. Scientifically if and when 64 millions of SNPs have been assumed to be mutational alleles, it is a valid question that how long it took to complete 64 million mutations and the YEC researchers have simply narrated their observation, carried out in the simulation experiment and they defined that they assume that at the rate of 100 mutations per person per generation the accumulation of 64 million mutations in 7 billion human population (the current human population on Earth), it would take less than one generation (Carter and Hardy, 2015; Buggs, October 28, 2017; Sanford et al, 2018). It has further been simulated that if the population size has been assumed to be restricted to 10,000 individuals, the accumulation of 64 million mutations would take 80 years of generations. Though population biologists have contemplated that the accumulation of mutational alleles is considerably less due to intermittent drifting out of genetic variations out of the gene-pool during the cascading mode of vertical growth of population size, through generations. But the matter of fact is that the intermittent genetic drifting could have been offset if the concerned population have the potential to grow at an exponential rate. So, the above-mentioned observations on the genetic variation of a smallest biological population, that has gone through a short span of genetic bottleneck and have a potential of exponential growth could have retained maximum of its archaic genetic diversity, in the pre-bottleneck era and if this observation is not scientifically incorrect, the possibility of recognizing Adam and Eve as a primordial couple and considering them as the progenitor of the modern, well genetically diversified and derived population of modern human species on Earth should not be considered as a baseless, utopian idea either. When it has broadly been accepted as a hypothetical assertion that Adam and Eve could have given rise to the genetically diverse modern human race, the supporter of YEC group has started facing a narrow bottleneck of perceptional opposition by the evolutionists who have started starting specific objection like "Adam and Eve could not possibly account for the specific patterns of allele frequencies that we see in the modern human population" and that has found to be difficult for YEC contenders to dismiss without upgrading the version of their first computer-based simulation program "Mendel" to make an upgraded version called "Mendel-Go" (Version 2.7.2), to make sure that program could deal with the experimental simulation of population size, which is dynamic and the changing of allele frequencies could be tracked down, across the generations (Mitra et al, 1999; Sanford et al, 2018). The overview of the YEC hypothesis as well as Intellectual design is very straight forward, that claims that Adam and Eve with enough internal genetic diversity, the primordial couple in Eden is the miraculous, holy creation of God and has been attributed, millions of designed SNPs, as part of the Designed Diversity Model and the modern human race have been inherited with identical allele frequency descended from the primordial couple.

To demonstrate the actual allele frequencies of the modern human race, the YEC scientists have engaged in the critical investigation on the sequenced data for chromosome number 22, Y chromosome, and the mitochondrial DNA of modern human species, where Variant Call Format or VCF format of existing data (stored and used in Bioinformatics of each gene), presenting the Allele Frequency or AF for each ALT Allele, have been transposed on Minor Allele Frequency or MAF Plots, that represents the distributional index of actual allele frequency of the contemporary human population (Sanford et al, 2018; Hössjer and Gauger, 2019). It is the quite intriguing observation made by the ID advocacy group of scientists and YEC contenders as the use the MAF plot data as a control to compare the numerical simulation data, available from Mendel-GO simulation run and it is interesting to notice that Mendel-Go has found to be competent to keep an accounting of incoming and outgoing virtual alleles of the virtual dynamic population and able to quantify genetic transactions of the concerned species, accumulated for several generations, tally it with controls, plotting it in MAF and finally yield the allele frequency distributions (Sanford et al, 2018; Sarsani et al, 2019). In determining the allele frequency distribution of modern Human species and a comparison of allele frequency of it in comparison to the allele frequency of its progenitor couple on Earth, the Mendel-Go numerical simulation model has found to be a unique numerical simulation model that has taken into consideration several key issues like the profile of its created allele, state of mutation accumulation, state of small founder population, growth of population, population bottleneck, regrowth after a population bottleneck, etc.

The numerical simulation, done by YEC scientists, amazed with the outcome of Mendel (Version 2.7.2) has found to deliver a clear narrative about creation and tracking of two different types of genetic variation when a derived form of species with dynamic population size has been descended out of the mother stock of a small founder population size (Sanford et al, 2007b; 2018). Mutations are always occurring evolutionary processes found in a biological organism, which has found to be increasing in the lapse of time. Out of these two different types of genetic variations, the erstwhile and consistent nature of genetic variations have been designated as designed alleles whereas the alleles, that analogically emerged as a result of "word-processing errors," considered as mutational alleles, which have been emerged as rare variants. It has been observed by the scientists that from the beginning, Mendel is the numerical simulation program that keeps tracking each unique mutation and every mutational allele as well (Sanford et al, 2007a; Sanford et al, 2018). Furthermore, these YEC scientists noticed that to simulate newly emerged mutational alleles, they need to feed the data of the population's mutation rate and relevant data on the effect of each mutation on fitness in the Mendel. According to a feasible review of the mutational rate of modern human species, scientists have determined that the number of mutations, occurred per person per generation is around 100 and its default mutational effect of is "near-zero" as there was no selection happening for any mutational alleles for the occurrence of evolution and that lead drifting of all mutational alleles (Sanford et al, 2018). Furthermore, in addition to mutational alleles, the YEC advocating geneticists noticed that their simulation run through Mendel (with a new computational function), can track the designed allelic pairs (found in the ratio of 1:1 or 1:3), having designed functions, its location and fitness effects and the designed allelic pairs have been recognized with equal but opposite fitness effects. According to the numerical simulation run, made by the YEC supporting group of geneticists, of the "*Heterozygous Adam and Eve Model*," there's supposed to be four copies of chromosomes, possessed by this primordial couple in Eden, so the ratio of the designed allelic pairs in each chromosome would be the ratio of 50/50 or 25/75 at the rate of allele frequencies for the designed alleles would likely to be 0.25 or 0.50 to 0.75 (Sanford et al, 2018). The main objective of simulating "*Heterozygous Adam and Eve Model*," was to find out if there were any parameter profile that renders a heterogeneous primordial couple to transcend identical allelic frequency, it possessed itself in early days and the scientists have contemplated further that as each of the 88 primordial autosomes, found in Eden, likely carried their own unique set of designed variants that supposedly gone through the genetic process of segregation, recombination and come up with a vast array of genotypes after getting in next-generation and following generations, where the chance of increasing of allelic diversity is there as a combined result of the accumulation of newly emerged mutations and random rate of recombination of allele variants (Sanford et al, 2018; Gauger, 26th November 2019). YEC scientists have tried to figure out whether special creation power of God had rendered designing of gametes (gametogonia) of the Primordial couple in the Eden, so each of the gamete (gametogonium) had been designed to have its unique genotype. The scientists have logically defined that the two pre-historic designed couples might have anthropologically been related to the millions of individually designed gametogonia, which resembles to a gene pool, equivalent to the contemporary gene pool of a fairly large human population. So, a computer-driven numerical simulation has been run on an upgraded version of "*Mendel Go*" simulated 50 descendants, carried designed alleles from the primordial couple that would have transcended through 100 genetically independent gametes (comprised of 50 sperms and 50 eggs) (Sanford et al, 2018). In course of running this simulation through Mendel, an account of initially designed alleles and accumulating mutational alleles have been maintained, following the Biblical narrative of 200-generation, having sub-cycles of growth of erstwhile population, an incident of the 6-person bottleneck (in Noa's Ark, during Genesis flood), instead of population growth in post-flood bottleneck to a "pre-set population size" (Sanford et al, 2018; Buggs, 4th March 2020).

The geneticists have construed that alleles are the outcome of random mutations, so technically there are two different types of alleles have been recognized, ancestral or original allele and the derived or mutant allele and the geneticists further considered that the original allele is most common, so it has been recognized as major allele and the mutant allele is the rare one, should be recognized as a minor allele, which has been found in the allele frequency plot, but the major alleles have remained in empirical genetical reviews (Sanford et al, 2018). At the time of reviewing the human allele frequency distribution data, the geneticists failed to distinguish whether the allele under their investigation were original or derived in nature due to the dynamic nature of allele frequencies, but the YEC hypothesis supporting scientists claimed that their numerical simulation has rendered distinction between the original alleles and the derived one, emerged as a result of mutation. As the minor alleles are found to be visible on the allele frequency plot, the geneticists have contemplated the frequency of it is f, so the frequency of a major allele is $(1-f)$. The geneticists further observed that if the distribution of the major alleles would have been plotted in the allele frequency plot, it would supposedly for a mirror image of minor allele distribution, making the figure of allele frequency unessential. It is little tricky but geneticists have claimed that for investigating designed alleles a comparative account of all types of alleles (either major or minor) needs to be done, but it has mostly presented in form of allele frequency plot with minor alleles (1-50%) unless it requires an exceptional format of allele frequency plot with an exclusive presentation of minor and major alleles on the composite allele frequency plot (Keinan and Clark, 2012; Buggs, April 14, 2018; Sanford et al, 2018). YEC supporting geneticists have observed that in their simulation experiment with designed alleles, it's difficult to distinguish the ancestral or major and derived or minor alleles as the frequency of both categories of contrasting alleles be around 50% but they mentioned that the occurrence of genetic drifting has amicably settled down the account as scientists have critically noticed that as long as genetic drift continue its function, the flipping of major and minor allele have taken place and the empirical existence of "less abundant alleles" have rendered the scientists to recognize it as minor alleles (Greenbaum et al, 2014; Sanford et al, 2018). To run their numerical simulations of evolutionary human populations, the scientists have chosen evolutionary human populations (of 1000 individuals) having no designed alleles and the scientists inducted 100 new mutations to each individual per generation (Sanford et al, 2018) and they have restricted the nature of mutations in a neutral state to make sure the mutational variations would have to cascade through several generations as the nature of deleterious alleles would have pushed the entire population in extinction. In their designed allele simulations, these scientists have chosen the contrasting allele pair, the sum of allele frequencies should be 1.0 and both of these alleles, supposedly stationed at the same linkage block, each of which is identified by the alternative nucleotides at the same genetic locus (Sanford et al, 2018). Furthermore, these geneticists have considered all designed alleles co-dominant in running simulation and track the part of the population, that are initially heterozygous. If the fraction has emerged as 1.0, the scientists have conveniently defined the result as designating Adam and Eve as heterozygous, where the entire designed alleles supposedly initiated with the frequency of 50%, then the scientists have tried to tender an alternative interpretation of the fractional index of 0.5, indicating either Adam or Eve as heterozygous for entire alleles and it rendered identifying another one as homozygous and in this alternative scenario, the designed alleles have initiated at a frequency of 25% or 75%. However, the allele frequency distributions as per 1000 Genomes project data (2015) for Chromosome 22, chromosome Y and MT have been examined under three scenarios (Faulk, 2004; Carter, 2015; Gauger, October 3, 2017; Baharian and Gravel, 2018; Sanford et al, 2018):

Scenario1: The minor allele frequency distribution for human chromosome 22 had been simulated, where the number of sequenced individuals was 2504, the cumulative number of SNPs found to be around 918,038 and the common SNPs were 215,313.

Scenario 2: The allele frequency distribution for human chromosome Y had been simulated, where the number of sequenced individuals was 1233, the cumulative number of SNPs was 60,446 and the common SNPs were 7491.

Scenario 3: The allele frequency distribution for the human mitochondrial chromosome had been simulated, where the number of sequenced individuals was 1074, the cumulative number of SNPs was 2618 and the common SNPs were 424.

After careful review of the three scenarios, mentioned above, the YEC contending scientists have decided to set the first scenario as the yardstick, hence the allele frequency distribution of chromosome 22 for human autosomal chromosomes as the feasible proxy for the rest of the genome (Sanford et al, 2018). During running their numerical simulation, the scientists observed that the mutational allele frequency distributions have been influenced by the rate of genetic drift, and the influence of variables like the size of the population, rate of mutation, and the time influence in determining the rate of genetic drift. It has generally been observed that in any evolutionary simulation experiment about any biological population irrespective of its location and size, the accumulated mutations have found to build up linearly over time. It has also been noticed critically that the mutational alleles have intruded in any population and drifted the opposite direction of genetic drift (away from zero), slowly. Furthermore, the rate of genetic drift in a population around 1000 or more found to be very slow. When a population has accumulated mutations for a time being, getting big enough to reach a state of equilibrium between mutation and equilibrium, as the new alleles drift in the population, the older alleles started counterbalancing it by getting out of that population and in this state, the stabilization of allele frequency distribution and the hold of polymorphism increasing has been witnessed with four possible scenarios (Carter et al, 2008; 1000 Genome project,2015; Hössjer et al, 2016; Sanford et al, 2018):

Scenario A: Hence, in a simulation run, where the population size had been restricted to 1000, where the mutation rate was around 100 mutations per individual per generation and where is run for 10,000 generations, and it represents an allele frequency from 1-100%, a state of mutation/drift equilibrium has been witnessed, where fixation of a large part of allele happened at a high rate.

Scenario B: It follows an identical simulation path of Scenario A but it involves the minor allele frequencies (1-50%). The scientists considered Scenario B as the yardstick for portraying a stable evolutionary allele distribution. Furthermore, it has been observed that this second category of simulation of the evolutionary population of time being yield an allele frequency distribution that has found to be different from the 1000 genomes project (2015). If a comparative graphical scenario is concerned, the observed allele distribution has found to be "tighter bend" in comparison to the evolutionary scenario, at the frequency range of 3-15% (Sanford et al, 2018).

Scenario C: In this evolutionary simulation run in the Mendel-Go program, an effort has been to determine the frequency allele of the primordial couple, i.e. Adam and Eve scenario. A similar numerical simulation has been done as same as Scenario B, except a bottleneck that occurs 200 generations before (i.e. 10,000-200 = 9,800 generations) the run is finished, subsequently inducted a surge of population growth at the rate of 1.5 immediately after bottleneck to recover the shrinking population size. So, the simulation run analogically represents a scenario where a bigger population (which is comprised of 1000 individuals) shrinks to two, followed by a surge of it to replenish in its previous state. So, it gave an analogical interpretation where it shows that evolutionary Adam and Eve historically hold a transitional position in this evolutionary trail, where this primordial couple has been transcended from a larger human population in the past and give rise to the modern human population nowadays. The scientists defined that the allele distribution of Evolutionary Adam and Eve has found to be matching to the Scenario 1 (which has been defined earlier) but it is found to be distinct from Scenario B as graphically the bend of the curve is found to be more sharp and distinct in this simulation, that represents a rate of fixation is much consistent and sets in earlier in simulation run (Sanford et al, 2018).

Scenario D: In this simulation run, the initial set up of the size of the population is found to be exactly like scenario A, and the scientists tried to see the effect of mutation accumulation, that complies Biblical time frame and population dynamics, where the run starts with two individuals followed by a high rate of population growth till 10th generation followed by a bottleneck that brought down the population size to six, subsequently followed by a high rate of population growth that helped to replenish the population size to its erstwhile state, that is 1000 individuals and the runs end in 200 generations (Sanford et al, 2018). This entire simulation run does not fit the allele distribution model of the observed of the autosomal profile of chromosome 22 (i.e. Scenario 1), rather it fits the allele distributions of chromosome Y of Scenario 2 and Mitochondrial chromosome of Scenario 3. The YEC supporting geneticists further defined that from the perspective of the special creation of Biblical narrative, it's not a void perception as it has been contemplated that in Eden four copies of autosomal chromosomes have supposedly been existed, rendered initial state of heterozygosity, where there have supposedly existed a Y chromosome and a Mitochondrial chromosome (Underhill and Kivisild, 2007; Sanford et al, 2018). So, the erstwhile four copies of autosomes, have been considered to be responsible for designed heterozygosity, where the status of ChrY and Chr Mit were assumed to be invariants and that elucidates further that the erstwhile Edenic (non-variant) chromosomes were lack of designed variants and the geneticists tried to support their hypothesis by referring the presence of few numbers of SNP alleles, and the low number of allele frequencies and such interpretation has found to be a feasible explanation when the chromosomes were found to be young and initially invariant so it has substantiated further that the autosomal chromosomes, those that lack of designed variants, could be valid interpretation if those supposedly appeared at very young state in historic time, as per Biblical interpretation of time regime (Smolenyak and Turner, 2004; Underhill and Kivisild, 2007).

After traversing through a long trail of the evolutionary labyrinth, the geneticists are looking for feasible answers of an enigmatic issue whether the allele frequency distributions, witnessed among modern human species, could be derived from the primordial nature of heterozygous Adam and Eve and the scientists have engaged in running distinct options of numerical simulations and try to find out feasible options needed to be elaborated under following heads (Yuan et al, 2012; Buggs, October 28, 2017; Sanford et al, 2018; Hössjer and Gauger, 2019):

A. Running simulations with designed alleles (having an initial frequency of 50%): The scientists have ran simulations with designed alleles and ensured that no emergence of new mutations interfere and it has been initiated with all alleles with a frequency level of 50%, whereas most of the experimental simulations, pertaining with allelic frequency deal with minor alleles with a frequency range between 0-100%. For running this specific simulation, scientists have adopted several preset conditions like a short-span of running restricted to 200 generations, number of allele pairs restricted to 989, number of linkage blocks engaged with 989, no mutations and no selection effect (which is technically recognized as nearly-neutral fitness effect).
   The graphical presentation of this numerical simulation just amazed the scientists as they noticed the allele distribution in form of a straight spike at 50% (first generation) graphically representing the initial allele frequency (i.e. SNP frequency) distribution, where the population size is restricted with 1000 individuals and the total period restricted to 200 generations. At the end of 200 generations, the two-dimensional graphical column had taken a bell-shaped formation that indicated that there was little or no occurrence of genetic drift in this phase of the simulation. In the next phase, a pro-Biblical influenced simulation had been run, involving identical allele frequcncy, with a population size of two, with a time of 200 generations, followed by a steady surge of population, subsequently followed by a having a population bottleneck to 6 people in the middle on the 10th generation and ended up with a rapid population surge to get back to its early population size (i.e. 1000) and the scientists noticed that a couple of population constrictions could yield in extensive allele frequency spreading.

B. Running simulations with designed alleles (having an initial frequency of 25%): In this phase of the simulation experiment, scientists simulate with designed alleles specifically with the minor allele frequency of 25% level. It had been observed by the scientists that for each pair of alleles, where the frequency of minor allele pair is 25%, the contrasting allele with an allelic frequency would be 75%. After simulating 20 generations, scientists noticed the two-dimensional graph would have appeared like bell-shaped curve with a spike at 25% but it would be looking different after 200 generations as the bell-shaped curve would become straight then get straight-line slope at the right hand which is considered to be the actual allele frequency distribution. However, the YEC supporting geneticists had recognized such low-level initial allele frequency, which had been observed in the simulation model, its convenience to fit in the "observed human frequency distribution," elucidated earlier in Scenario 1. The geneticists had further observed that the nature of the slope line had been altered if a fraction of allele pairs (those begin at 50% level, rather than 25% level) included, moreover inclusion of mutation alleles in greater extent in the simulations rendered diminishing the steepness of the slope.

C. Simulating with high-impact designed alleles (emerged as a result of selective sweeps): In this simulation experiment, geneticists have dealt with a small number of the designed allele in pairs (initially at 50% level), that were assigned broad fitness effect and fit in Biblical narrative. The designed alleles were of composite type, comprised of minor or less favored and major or more favored alleles. The scientists, engaged in running this simulation, critically observed that most of the allele pairs had responded in natural selection, moved sideward in the extreme left and right flank, leaving the central part empty (forming Gap zone) to get a shape of H frame curve and this simulation run has conveniently segregated "high frequency created alleles" and "low-frequency mutational alleles" (Sanford et al, 2018). The geneticists, supporting the YEC hypothesis, had contended further that a series of selective sweeps have the potential to mask the effect of genetic drifting.

D. Running simulations with designed and mutational alleles: In this simulation experiment, the scientists had combined designed alleles (initially at the level of 50%) with mutational alleles helped to yield the resulting the left-ward movement of rare mutational alleles on the x-axis and it did not fit in the observed allele distribution of Scenario 1. The geneticists carried out this simulation experiment, in compliance with the biblical parameters, like the growth of population starts from the primordial couple, followed by ten generations of population growth, subsequently followed by a single generation bottleneck, resulting in a shrinkage of the population to 6 individuals. The scientists noticed a spike of alleles in their graph at the far left, and that had supposed to have occurred as a result of the accumulation of mutational alleles throughout the simulation (i.e. for 200 generations)

   In the next phase of a simulation run, which was popularly recognized as a "*Heterozygous Adam and Eve Simulation*," had been run in compliance to the biblical parameters, where accumulating mutational alleles had been put together with the initially designed alleles (initially all at 25%) in the simulation run (Buggs, October 28, 2017). It needs to be mentioned here that the geneticists had picked 296,700 initial designed alleles comprised of a big chunk of less favored alleles and lesser quantity of more favored alleles where results shows that the rare mutational alleles moved left-ward, resulting in the spread out of designed alleles along the X-axis and the resulting allele frequencies plotted by Mendel-Go version, from 1-50% range, had found to be identical to the observed allele distribution of Scenario 1. The scientists had critically observed that the shape of the curve depended on the population growth rate of early population and various aspects of the population dynamics like-selective sweeping, extinction of lineages, etc.

In absence of direct geological, anthropological and paleontological evidence and wide-spread nurturing of enigmatic molecular foot-print by molecular geneticists and genomicists since the late nineties, scientists have engaged in a diverse range of experiments and tried to harness each and possible means of investigations, including numerical simulation to have an idea of genetic diversity, encrypted in the genome (the ultimate holy-grail of life) of the primordial couple (Adam and Eve), by making an appraisal on the variation of designed alleles and the preliminary investigation on the low-frequency alleles by running numerical simulation have helped the geneticists to analyze the allele distribution patterns of modern human. So, the YEC supporting geneticists have claimed with the firm affirmation "*If we start with the premise of a miraculously created Adam and Eve, the idea of 'designed diversity' is a logical deduction. It provides the most coherent explanation for the beneficial variations that we see within human race today. Since all parties acknowledge that nearly all non-neutral mutations are deleterious, only designed variants can credibly account for all the 'good diversity' we see (i.e. variations that are desirable, and have no pathological effects)*" (Sanford et al, 2018). The YEC group of scientists strongly expressed that they don't believe that the desirable human variations can not be attributed to the stereotypic Darwinian evolutionary templates of mutation/selection (Lacy, 1987; Orr, 2007; Loewe, 2008). After a thorough review of a wide array of results generated by a series of the simulative run, the scientists, supporting the hypothetical emergence of humanity out of the cradle of designed diversity, strongly claimed that if and when designed diversity is coupled with natural selection, it triggers the extent and magnitude of instantaneous local adaptation and the multiple disjunctions of the main population to yield a series of sub-populations, that happened after the bottleneck of Genesis Flood (Lacy, 1987; Carter and Powell, 2016). Though the evolutionists would never agree to the perception of the special creation of humanity as they strongly supported that evolution of any living entity has been integrated to nature rather than considered to be emerged out or integrated to any special power of special creation, the YEC contending geneticists relied upon the emergence of humans and/or other living entities on Earth had strictly integrated to the perception of "designed diversity," and the emergence of humans or the journey of the humanity of Earth, has been contemplated as the part of God's design, that has a sacred inheritance of the erstwhile couple on Earth, specifically starting with Adam and Eve in Eden on the Earth in the pre-historic time of Bible.

References

Ayala, F. J. (1995). The myth of Eve: molecular biology and human origins. *Science.* 270(5244): 1930-6. doi: 10.1126/science.270.5244.1930.

Ayala, F. J., and Escalante, A. A. (1996). The evolution of human populations: a molecular perspective. *Mol Phylogenet Evol.* 5(1):188-201. doi: 10.1006/mpev.1996.0013.

Ayala, F.J., Escanlante, A., O'hUigin, C. and Klein, J. (1994). Molecular genetics of speciation and human origins. *PNAS USA* 91: 6787-6794.

Baharian, S., and Gravel, S. (2018). On the decidability of population size histories from finite allele frequency spectra. *Theor Popul Biol.* 120:42-51. doi: 10.1016/j.tpb.2017.12.008.

Baumgardner, J., Brewer, W., and Sanford, J. C. (2013). Can synergistic epistasis halt mutation accumulation? Results from numerical simulation. In: Marks RJ, Behe MJ, Dembski WA, Gordon B, Sanford JC (eds) *Biological information – new perspectives*. World Scientific, Ithaca, pp 338–368, 312– 337.

Buggs, R. (October 28, 2017). Adam and Eve: a tested hypothesis? https://natureecoevocommunity.nature.com/posts/22075-adam-and-eve-a-tested-hypothesis

Buggs, R. (April 14, 2018). Adam and Eve: lessons learned. http://richardbuggs.com/2018/04/18/adam-and-eve-lessons-learned/

Buggs, R. (March 4, 2020). Adam and Eve our ghostly ancestors? https://natureecoevocommunity.nature.com/posts/61200-adam-and-eve-our-ghostly-ancestors?channel_id=522-journal-club

Carter, R., Criswell, D., and Sanford, J.C. (2008). The "Eve" mitochondrial consensus sequence. In *Proceedings of the Sixth International Conference on Creationism*, ed. A.A. Snelling, pp.111–116. Pittsburgh, Pennsylvania: Creation Science Fellowship and Dallas, Texas: Institute for Creation Research.

Carter, R. W. (2015). Finally setting the record straight—Christian views on the historicity of Adam through the ages. Journal of Creation 29(2):36-40.

Carter, R.W. (2017). Reading evolution into the Scriptures: a review of Adam and the Genome: Reading scripture after genetic science (Dennis R. Venema and Scot McKnight). *Journal of Creation* 31(2):41–46.

Carter, R.W. (2018). Effective population sizes and loss of diversity during the Flood bottleneck. *Journal of Creation* 32(2):124–127.

Carter, R. and Hardy, C. (2015). Modelling biblical human population growth. *Journal of Creation* 29(1):72–79.

Carter R., and Powell, M. (2016). The genetic effects of the population bottleneck associated with the Genesis Flood. *Journal of Creation* 30(2):102–111.

Faulk, D. (2004). *Coming to peace with science: Bridging the worlds between faith and biology*. Intervarsity Press, Downers Grove, IL.

Gauger, A. (October 3, 2017). Does science rule out a first human pair? Geneticist Richard Buggs says no. Retrieved March 22, 2018 from http:// www.evolutionnews.org/2017/10/does-science-rule-out-a-first-humanpair- geneticist-richard-buggs-says-no.

Gauger, A. (November 26, 2019). Human origins: Not a simple question. https://evolutionnews.org/2019/11/human-origins-not-a-simple-question/

Greenbaum, G., Templeton, A.R., Zarmi, Y., and Bar-David, S. (2014). Allelic richness following population founding events--a stochastic modeling framework incorporating gene flow and genetic drift. *PLoS One*. 2014 Dec 19;9(12):e115203. doi: 10.1371/journal.pone.0115203.

Harpak, A., Bhaskar, A., and Pritchard, J.K. (2016). Mutation Rate Variation is a Primary Determinant of the Distribution of Allele Frequencies in Humans. *PLoS Genet*. 12(12):e1006489. doi: 10.1371/journal.pgen.1006489.

Hossjer, O. and Gauger, A. (2019). A single-couple human origin is possible. *BIO-Complexity.org* 2019(1): 1-20.

Hössjer, O., Gauger, A., and Reeves, C. (2016b). Genetic modeling of human history part 2: A unique origin algorithm. *BIO-Complexity* 2016(4): 1-36.doi:10.5048/BIO-C.2016.4.

Jeanson, N., and J. Tomkins. (2017). Genetics confirms the recent supernatural creation of Adam and Eve. In: *Searching for Adam*, ed. T. Mortenson, pp. 307–312. Green Forest, Arkansas: Master Books.

Keinan, A. and Clark, A. G. (2012). Recent Explosive Human Population Growth Has Resulted in an Excess of Rare Genetic Variants. *Science* 336 (6082): 740-743.

Lacy, R. C. (1987). Loss of genetic diversity from managed populations: Interacting effect of drift, mutation, immigration, selection and population subdivision. *Conservation biology* 1(2): 143-158.

Levy, S., Sutton, G., Ng, P.C., Feuk, L., Halpern, A.L., Walenz, B.P., Axelrod, N., Huang, J., Kirkness, E.F., Denisov, G., Lin, Y., MacDonald, J.R.,Pang, A.W. C., Shago, M., Stockwell, T.B., Tsiamouri, A., Bafna, V., Bansal, V., Kravitz, S.A., Busam, D.A., Beeson, K.Y., McIntosh, T.C., Remington, K.A., Abril, J.F., Gill, J., Borman, j., Rogers, Y-H., Frazier, M.E., Scherer, S.W., Strausberg, R.L., and Venter, J.C. (2007). The diploid genome sequence of an individual human. *PLoS Biology* 5(10):e254.

Loewe, L. (2008). Negative selection. *Nature Education* 1(1):59.

Mitra, A., Yadav, B.R., Ganai, N.A., and balakrishnan, C.R. (1999). Molecular markers and their applications in livestock improvement. *Current Science* 77(8): 1045-1053.

Myers, S., Fefferman, C., and Patterson, N. (2008). Can one learn history from the allelic spectrum? *Theor Popul Biol.* 73(3):342-8. doi: 10.1016/j.tpb. 2008.01.001.

Nei, W., Maruyama, T., and Chakraborty, R. (1975). The bottleneck effect and genetic variability of populations. *Evolution* 29(1):1–10.

Orr, H.A. (2007). The probability of parallel evolution. *Evolution* 59(1): 216-220.

https://doi.org/10.1111/j.0014-3820.2005.tb00907.x

Rupe, C.L., and Sanford, J. C. (2013). Using numerical simulation to better understand fixation rates, and establishment of a new principle: Haldane's Ratchet. In Proceedings of the *Seventh International Conference on Creationism*, ed. M.F. Horstemeyer. Pittsburgh, Pennsylvania: Creation Science Fellowship.

Sanford, J.C., Baumgardner, J., Gibson, P., Brewer, W., and ReMine, W. (2007a). Mendel's accountant: a biologically realistic forward-time population genetics program. *Scalable Comput: Pract Exper* 8(2):147–165.

Sanford, J.C., Baumgardner, J., Gibson, P., Brewer, W., and ReMine, W. (2007b). Using computer simulation to understand mutation accumulation dynamics and genetic load. In: Shi Y, van Albada GD, Dongarra J, Sloot PM (eds) *International conference on computational science*. Springer, Berlin/Heidelberg, pp 386–392.

Sanford, J.C., and R. Carter. (2014). In light of genetics…Adam, Eve, and the creation/fall. *Christian Apologetics Journal* 12(2):51–72.

Sanford, J., Carter, R., Brewer, W., Baumgardner, J., B. Potter, and Potter, J. (2018). Adam and Eve, designed diversity, and allele frequencies. In *Proceedings of the Eighth International Conference on Creationism*, ed. J.H. Whitmore, pp. 200–216. Pittsburgh, Pennsylvania: Creation Science Fellowship.

Sanford, J., and Nelson, C. (2012). The next step in understanding population dynamics: comprehensive numerical simulation. In: Carmen Fustú M (ed) *Studies in population genetics*. ISBN: 978-953-51-0588-6 InTech.

https://www.intechopen.com/books/studies-in-population- genetics/the-next-step-in-understanding-population-dynamics- comprehensive-numerical-simulation

Sanford, J.C., Brewer, W., Smith, F., and Baumgardner, J. (2015). The waiting time problem in a model Hominin population. *Theor Biol Med Model* 12(18). https://doi.org/10.1186/s12976-015-0016-z

Sarsani, V. K., Raghupathy, N., Fiddes, I. T., Armstrong, J., Thibaud-Nissen, F., Zinder, O., Bolisetty, M., Howe, K., Hinerfeld, D., Ruan, X., Rowe, L.,

Barter, M., Ananda, G., Paten, B., Weinstock, G. M., Churchill, G. A., Wiles, M. V., Schneider, V. A., Srivastava, A., and Reinholdt, L. G. (2019). The Genome of C57BL/6J "Eve," the Mother of the Laboratory Mouse Genome Reference Strain. *G3 (Bethesda)*. 9(6):1795-1805. doi: 10.1534/g3.119.400071.

Shaffner, S. (2017a). Can someone explain like I'm 5 yo, what's wrong with this refutation of Biologos? Retrieved May 15, 2018, from https://discourse.biologos.org/t/can-someone-explain-like-im-5-yo-whatswrong- with-this-refutation-of-biologos/5657/16 [See also post #126 on this same thread]

Shaffner, S., (2017b). Bottleneck: a forward genetic simulator for playing around with demographic bottlenecks. Retrieved May 15, 2018, from https://github.com/glipsnort/bottleneck [This was the software used to generate the graphic in the discussion thread (Shaffner 2017a)].

Smolenyak, M.S. and Turner, A. (2004). *Trace your roots with DNA: Using genetic tests to explore your family tree.* Rodale Books. P. 288.

The 1000 Genomes Project Consortium. (2015) A global reference for human genetic variation. *Nature* 526: 68–74. https://doi.org/10.1038/nature15393

VanDoodewaard, W. (2015). The Quest for the Historical Adam: Genesis, Hermeneutics, and Human Origins. Grand Rapids, Michigan: Reformation Heritage Books.

Venema, D.R. 2010. Genesis and the genome: genomics evidence for human-ape common ancestry and ancestral hominid population sizes. *Perspectives on Science and Christian Faith* 62(3):166–178.

Venema, D.R., and S. McKnight. (2017). *Adam and the Genome: Reading Scripture After Genetic Science*. Brazos Press, Grand Rapids, MI.

Wood, T.C., and J.W. Francis. (2016). Genetics of Adam. In: *What Happened in the Garden*?, ed. A. Chou, pp. 75–100. Grand Rapids, Michigan: Kregel.

Woodmorappe, J. (1996). *Noah's Ark: A Feasibility Study*. Inst for Creation Research. P.298.

Yuan, Q., Chu, C., and Jia, J. (2012). Association studies of 19 candidate SNPs with sporadic Alzheimer's disease in the North Chinese Han population. *Neurol Sci.* 33(5):1021-8. doi: 10.1007/s10072-011-0881-0.

Chapter 10

Testifying the History of Human Origin: Looking Through Spectacles of BTSC (Bottleneck to Single Couple) & SCO (Single Couple Origin) Model

Additional research indicates that there may be as few as three or four ancestral versions of the region surrounding the HLA-DRB1 gene, when all sequence comparisons are taken into account. This is tremendously exciting, since Ayala's [Francisco J. Ayala] claim of thirty-two ancestral alleles has frequently been used to argue that Adam and Eve cannot have existed. But with additional study, those claims have been shown to be false, and the genetic argument now no longer contradicts the possibility of two first parents. Three or four alleles can be carried by just two individuals, especially if after those two first parents came a period of rapid population expansion.

---- ***Ann Gauger***

Being a social animal, most of our perception about enemies and friends is very distinct as per stereotypic perception, we could easily recognize that person as the enemy who consciously opposed or showed hostility to someone or something and our inherent feelings help us to recognize a person as a friend with whom we have a mutual bond of affection or a strong sense of fraternity, but it happens so often to us that we don't clearly distinguish between these two in our professional life so that it makes us difficult to recognize friend or enemy in our professional arena. It is difficult to recognize our friend but it is not so difficult to recognize our enemy as in this regime of the rat race we have more enemies than a few numbers of friends in real life. To find out a true invisible friend, one of the best and alternate way is to take resort to the old proverb: "*The enemy of my enemy is my friend,*" which has been scripted in the great Sanskrit literature "*Arthashastra*" authored by Kautilya in 4th Century BC. The above statement has made it clear to accept that negative expressions made of any negatively expressed ideas or explanation, have helped to get positive impressions or images about something (some object, ideas, explanation, etc.,), some action or someone. Once the Evangelical Scientists have firmly adhered to their observation by discarding the theological perception as well as biblical narratives of the possibility of historical "Adam" as well as the traditional history of humanity by sharing their observations on popular websites like: www.npr.org/ 2011/08/09/138957812/evangelicals-question-the-existence-ofadam-and-eve. and www.christianitytoday.com/ct/2011/june /historical Adam.html., it not only created a big confusion among the church going communities, it passively instigates the atheists to fight against the supporters of special creationists to describe it as a moral win of science over religion.

The most interesting thing related to the evolution of modern human species out of ancient ape like animals is based on logical assumptions, rather than the presence of substantial evidence, observations and interpretations (Ayala and Escalante, 1996; de Groot et al, 2012). So, it naturally intrigues us to be inquisitive towards those articles, that strongly argued against the non-existence of Adam and Eve as the primordial ancestor of modern human species as we focus our attention on the nicely crafted logical arguments that refer the contemporary level of allele diversity of modern human species is too high to get through the bottleneck of two and the result of linkage disequilibrium has been referred to substantiate that the size of the ancestral population of 1000 or more individuals of the human population helped to yield the modern human species (Enns,2012; Gauger et al, 2012; Hossjer et al 2016a; Hossjer 2016b). So the logical interpretation of the emergence of modern human species out of the primordial couple of two seems to be holistic, which has little fact to accept the biblical narrative of genesis and it has logically seems to be tenable when scientists have stated: "*current DNA polymorphism (differences between DNA sequences between individuals) provides enough evidence to prove Adam and Eve never existed*" (Gauger et al, 2012). It is a matter of fact that the evolutionists have studied and examined the origin and evolution of modern human species, its genetic diversity, related to its past genetic history and this genetic diversity, ranges from a single base difference called SNPs or Single Nucleotide Polymorphisms or mobile genetic elements or variation of genes or allelic diversities, etc. (Rafajlovic et al, 2014). So, it must be taken into consideration that the models and equations that analyze and narrate the distribution of genetic variants (alleles) that present in the gene pool of a population must comply with a set of assumptions. During the review of these results and evolutionary models to judge the feasibility of the results and the efficiency of the model to upgrade or discard it or any comparative studies between two or in several models, the scientists have observed that some assumptions are presuppositions and a unique analogical reference has been tendered: "*the use of tree-drawing algorithms begs the question of whether or not a tree exists. This is an assumption drawn from Darwinism. Similarly, the assumption that random processes are the only causes of genetic change over time is an assumption drawn from philosophical naturalism. These assumptions may or may not justified*" (Gauger et al, 2012). The most important take-home message is that if a logical interpretation of any observation has been made on assumptions, the apparent sustainability of it depends upon the extent of justification of the assumptions and it might have been replaced by the emergence of an alternate nature of assumption that appears to be more justified. So, in the model-based interpretation of any scientific incident in the dynamic world of science, there is no everlasting nature of the findings, as "truth" of today have often been shifted to "virtual truth" of tomorrow.

In any investigations, related to population dynamics modelling, population geneticists have contemplated the studied population as "ideal population," which is random breeding in nature and the variables related to the population held constant during the experiment but it hardly happens in real life and there is no such population, regarded as "ideal population" in the biological world, so far, so it rendered Sewall Wright (1938) to introduce a newly defined terminology, called "effective population size" or "*Ne*" to indicate the size of a freely reproducing population, mating behaviour and the distribution of allele frequencies is identical to the actual size of population *N*. So, the population geneticists have used *Ne* in place of *N* to make the equation work but it's not easier to determine the *Ne* of any population as it is integrated to its reproductive biology and it's no easy task to know or finding out the reproductive behaviour of the ancestral population of any species (Gauger et al, 2012). However, it has been observed that most often the *Ne* has been used to challenge the existence of Adam and Eve. Let us go through a simple example to understand the importance of the genetic diversity of the modern human population in terms of understanding the functionality of allele frequency in it. It has generally been witnessed among the non-European community that digestion ability of milk among adult individuals is found to be very less, which has scientifically been recognized as lactose intolerance (Vuorisalo et al, 2012). An occurrence of a mutational change of lactase gene has happened in the Scandinavian communities in the evolutionary history of modern human species and has been gradually spread out in the global population rendered the big number of adult individuals able to regain the potential of digesting milk under carrying the lactase gene as well as a mutant allele of lactose resistant genes (Gauger et al, 2012). This change of gene mutation and spreading out of beneficial traits across the population influenced the global population to go for dairy farming. The occurrence of such beneficial mutation and spreading out of the mutant alleles in the communities depends on some factors like the size of the population, life-cycle of each generation or average time of each generation cycle, rate of mutation, extent of inbreeding in that population, reproductive success of such population in bearing and carrying forward the mutant alleles and the rate of migrations in and out of the concerned population, etc. (Nachman and Crowell, 2000). The lack of availability of any direct answer made it difficult for the geneticists to appraise the functional aspects of allele frequency and define the beneficial aspects of mutant alleles, witnessed in the population level. So it rendered the scientists to design a framework of estimation to review the sequence data by using SNP data to form an evolutionary tree for the DNA sequences of a particular gene and the further investigation of such trees along with the feasible assumption of the rate of mutation and predicted span of generation cycle, helped the evolutionary biologists to figure out "*at what point in time the trees coalesce to a single allele*" (Gauger et al, 2012). Once the geneticists have able to determine it, it would be much easier for them to determine the size of Ne, whenever the mutant allele first emerged out to define further the contemporary distribution of "*observed allele frequency of the lactase persistence gene*" (Gauger et al, 2012).

Let us find out the feasibility of estimation of the effective population size, where several assumptions have been made, like the assumptions on the consistent rate of mutation (as it has been perceived that there is no rate of variation witnessed in the DNA sequences overtime during the occurrence of mutation; but there is no reality in such contemplated assumption in real life (Gauger et al, 2012). In reality, the primary assumption of constancy of the population size is a void idea, as long as in and outward migration of such population happens intermittently. The geneticists have referred to another assumption, which has played a cardinal role in determining the effective population size that "*the DNA being analyzed is either not under selection, or there has been no change in a selection over time*" (Gauger et al, 2012). But that is not the reality as the molecular geneticists have noticed that in the case of lactase persistence allele, the strength of the allele is ever-changing rather than considered to be in the level of constancy (Gauger et al, 2012). The extreme level of advancement in the field of molecular genetics, have encouraged the scientists to experiment with linkage disequilibrium, which helped to estimate the tentative time of divergence of the effective population of the concerned species based on the calculation made on maiden emergence of any mutant allele like lactase persistence allele (Kimura and Ohta, 1971). According to the functional narrative of linkage equilibrium, the genes under selection travel together with closely linked genes in its proximity but this linkage between the genes concerned and the hitchhiking neighboring genes disappeared over time, depends upon recombination potential among them. The geneticists have mainly used this linkage disequilibrium information along with necessary information to determine the age of the mutant alleles, emerged the first time. The method of utilizing the linkage equilibrium has found to be a precise methodology to monitor the contemporary genetic as well as demographic changes of any population dynamics but the precise estimation and the feasibility of yield depend upon several assumptive factors like size of the population, selection factor, rate of recombination, rate of mutation, etc. (Gauger et al, 2012). It needs to be mentioned here that the initial level of findings by the scientists pertaining with linkage disequilibrium presented the tentative time of emergence of the lactase persistence allele, supposedly appeared in the modern human population somewhere in between 20,000 to 2,000 years ago, whereas an inclusive endeavor of fine-tuning of this primary result by including the analysis of "associated sequence of polymorphism" helped to get them a precise span of the emergence of this aforesaid allele, to project its time of emergence, likely took place between 12,000 to 7,000 years ago (Bersaiglieri et al, 2004; Coelho et al, 2005). So, the overlapping index of results helped the scientists to contemplate that the occurrence of mutation of lactose resistant gene or emergence of lactase persistent allele happened around 20,000 to 12,000 years ago in a broad sense and that has indirectly indicated a possibility that earliest effort of domestication of cattle happened around 20,000 years ago, but if we would like to narrow it down the gap of 7,000 to 2,000 years ago (which is around 5,000 years ago) has remained to be considered as a big gap and seems to consider some variables that would have helped the molecular genomicists to narrow down the gap (Itan et al, 2009; Gerbault et al, 2011).

The evolutionary biologists, especially the genome biologists, have referred the example of tentative emergence of a gene variant like lactase persistent allele, the tentative emergence of it have supposedly took place in between 20,000 years to 2,000 years ago and it made scientists argue that determining the allele diversity of the global population of modern human species and precise analyzing of the result, available from linkage disequilibrium statistics made the situation complicated to determine the exact time of emergence of the modern human population and its ancestral size, which the modern human race emerged from (Gauger et al, 2012). As choosing a DNA sample for sequencing seems to presently be a cumbersome process as it's difficult to choose the particular location of the genome as the structural and functional features of coding and non-coding region of genomes seems to be remarkably distinct, and selection of either one might have influenced in the varied projection of the ultimate result; moreover, the genome biologists observed after completion of the human genome mapping project that the earlier observation on the non-coding region of the genome, which has been contemplated as functionally inert region, found to be important location, engaged in a series of regulatory function nowadays (Gauger et al, 2012). The meticulous review on the constancy of the rate of selection has been found a vague idea as long as micro-evolution of modern human species is concerned as the rate of selection has found to be in the mode of acceleration in the evolutionary history of modern human species in last 10,000 years, that rendered the discarding of a constant rate of selection hypothesis (Hawks et al, 2007). The scientists have noticed that "genetic drift" (which expedite downgrading of allele frequencies) and "allele surfing" (that expedite offsetting of genetic erosion by triggering a surge of the population by enhancing the level of rare variants of gene immediately after genetic bottleneck) are found to be two counter-acting effects that preserve polymorphism and expedite the surge of the population after bottleneck so the extent of genetic erosion would have been replenished (Gauger et al, 2012). As it's difficult for the population biologists to deal with any population model where the size of the population is not constant and in reality, there is no such existence of static population where immigration/invasion is not there. The paleoanthropologists and the population biologists have observed that the modern human species have gone through an exponential surge of the population immediately after the bottleneck along with contemporary migratory expansion to occupy all congenial niches to occupy across the globe even the scientists have argued in favor of cross-breeding of the modern line of human populations with ancient lineage (like cross-breeding between Denisovans with Neanderthals) leading to the formation of genetic introgression (Quote: No 'anatomically modern' elephants, http://johnhawks.net/weblog/ ; Hawks, March 3, 2020). So, the matter of fact is that any genetically diversified large population without any issue of migration (in or outward) is likely to be a utopian assumption. Hence a level of inconsistency is always there when different population models have been used to determine the effective size of the population of modern human species during its very beginning of a journey, down the trail of evolution. In 2007, Albert Tenesa and his research colleagues have determined the effective size of the population based on linkage disequilibrium is found to range from 1,000 to 2,000 and with further bottleneck, it might have been restricted to even 100. As an example, a contemporary work of the population biologists needs to be mentioned here that has been done on the model of multiple genetic loci of modern human species, published in 2011 and it has mentioned the effective population size is around 14,000 (ranging from 2,000 to 20,000) (Blum and Jakobsson, 2011). So, it has been found that the size of the effective population of modern human species varies, depending on the nature of the population model, have been undertaken by the scientists to look for "correct" estimation and the nature of assumption, they made to determine a feasible result.

In 1987, scientist A.C. Wilson along with his research colleagues, R. L. Cann and M. Stoneking have analyzed the restriction mapping of the randomly drawn mitochondrial DNA sample of 147 people of the five geographical population across the globe and determined that all of the mitochondrial DNAs (Mitochondrial DNA has found to be descended from mother to the child through maternal inheritance) diverged out of a single woman of African origin, who has been contemplated to be alive 20,000 years ago, recognized to be as Mitochondrial Eve. In the same way, by sequencing the Y-chromosomes (that has exclusively possessed by men), the patrilineal inheritance of Y-chromosome from a single male of African origin, recognized as Y chromosomal Adam or Y chromosomal Most Recent Common Ancestor Adam or *Y-MRCA* Adam (Wilson Sayres et al,2014; Carter et al, 2018). This story began with Albert Perry, an African American, lived in South Carolina, USA, one of his female relatives had submitted his DNA sample to a commercial DNA analyzing lab called "Family Tree DNA" for genealogical analysis. Thousands of thousands of people get their DNA tested and cumulative analysis of the compiled data suggested that all males have got their Y-chromosome from a common male ancestor, recognized as "*Y-MRCA* Adam," who has postulated to be lived between 60,000 to 140,000 years ago (Barras, March 6, 2013). The earnest endeavor of the DNA techniques failed to place Albert Perry's DNA on the Y-chromosomal family tree and their meticulous studies of the possessor of this extraordinary Y-chromosome by Albert Perry has rendered them to contemplate that his male lineage has supposedly separated from the erstwhile ancestral archaic human lineage around 338,000 years ago (Barras, March 6, 2013). So these findings have intrigued the eminent geneticists from North America like Michael Hammer, Chris-Tyler Smith, and Jon Wilkins, etc. to observe that the Y-chromosomal tree is much older, beyond our conventional perception of its emergence between 60,000 to 140,000 years ago. The possible explanation of such existence of a historically old Y-chromosomal entity has been assumed to have emerged as a result of interbreeding of modern human species with the archaic humans like Neanderthals in the Middle East, or Denisovans in South East Asia (Barras, March 6, 2013). But a group of evolutionary biologists has strongly argued that the above-mentioned observation did not necessarily affirm that we the contemporary representative of modern human species, evolved out of the primordial pair of Adam and Eve. In 1995, Francisco Ayala, the eminent population biologist has challenged the existence of Adam and Eve as a primordial couple of humanity, as he examined the alleles of the gene, recognized as *HLA-DRB1*, many variants existed in the primates and humans and this gene has reported being responsible for yielding blood-cell protein engaged in our immune defense (Titus-Trachtenberg, 1994; Bergstrom et al, 1998; von Salome et al, 2007). The main argument of Ayala about the genetics of *HLA-DRB1* of the major histocompatibility complex entails a fairly large human population, supposedly evolved millions of years ago

(Ayala et al, 1994; Ayala, 1995; Ayala and Escalante, 1996). Several Biologos geneticists have the similar perceptions, Dennis Venema has found to be one of the leading ones who has claimed that the human evolution has been started with the archaic human population of thousands or more and he claimed that its certainty would have undoubtedly been compared to the act of heliocentrism (Venema and McKnight, 2017). However, the critical investigations on the genealogical studies of *HLA-DRB1*, made by Ayala (1995) has primarily ascertained that there are 32 variants in the ancestral population of human species, that helped to the divergence of chimpanzees and modern human species in course of evolution and the size of the erstwhile population of the ancestral human population, which has postulated to be few thousand. But the critical review of the published work of Ayala (1995) has ascertained further that the DNA sequencing examination, made on the *HLA-DRB1* gene had yielded an overestimated result regarding the determination of the size of the archaic human population and *HLA-DRB1* gene is found to be a variable gene in the human genome and it has a strong selection for heterozygosity, that supposedly yield a result, which is not found to be feasible, apparently (Kimura, 1969; Dorak et al, 2002; Kaimen-Maciel, 2009). The critical investigation of Bergsrtom et al (1998) on *HLA-DRB1* gene has indicated further that Ayala has studied the portion of the gene, found to be the hotspot region of the gene conversion (when the occurrence of mutation happen to be in accelerated rate) but the scientists under the leadership of Bergstrom (1998) have chosen to examine the same gene in a different part of it, where the rate of mutation is found to be at the baseline and it helped the scientists to explain that around 7 versions of gene existed in the ancestral population around 5 million years ago, during the divergence of hominids and great apes, taken place. During the emergence of modern human species, around 13 versions of the gene of *HLA-DRB1* existed in the ancestral population (Gauger et al, 2012). It has found to be more than a thousand variants of the *HLA-DRB1* gene in the present population of modern human species across the globe (www.ebi.ac.uk/imgt/hla/stats.html, accessed October 19, 2011). So the evolutionary and molecular biologists have contended that for genealogical studies, *HLA-DRB1* is not the right choice to review/examine an ancient genetic event as this gene is found to be under strong selection for increasing diversity of modern human species (Gauger et al, 2012). So Ayala's (1995) genealogical findings on *HLA-DRB1* gene with the existence of 32 alleles and nonexistence of Adam and Eve has found to be a credible story anymore in depicting the evolutionary journey of humanity, rather the contemporary findings of the scientists after examining the *HLA-DRB1*, claimed that "*there may be as few as three or four ancestral versions of the region surrounding the HLA-DRB1 gene, when all sequence comparisons are taken into account*" (Doxiadis et al, 2008). The possession of 32 ancestral alleles by a primordial couple (Adam and Eve) might be a quite unlikely but possession of 3 to 4 ancestral alleles by these primordial alleles seems to be a credible explanation of their possible existence if a surge of the population has supposedly happened after that (Gauger et al , 2012).

From the discussion of such data-driven conclusions, where several assumptions have been made for convenience and misinterpretation of the results have also been made, rather than intended to find out the inconvincible truth, that might have given an apparent edge to the science over religion, have been nurtured without making a sincere effort to examine the "single-couple origin hypothesis" (Hossjer and Gauger, 2019). However, in 2019, Ola Hossjer, a European mathematician from Stockholm University, Sweden and Ann Gauger, an eminent North American biologist from Biological Institute, Washington, USA have published their observation after comparing the results to allele frequency and linkage disequilibrium or LD findings is based on the numerical simulation run on the "Single-couple origin model" as they have engaged in secondary researches for few years on the published data on backward simulations, to gather previously published data to run the simulation (Yuan et al, 2012). It seems that research work of Ola Hossjer and Ann Gauger delighted the religious groups as the scientists tried to substantiate that journey of humanity started from Adam and Eve, the primordial couple at Eden as it supports the biblical narrative of the historical emergence of humanity on the Earth but it has been found that the radical religious followers of Abrahamic faith do express their skepticism about the main objective of this investigation and in an exclusive interview with Justin Taylor (October 28, 2019), Ann Gauger has made a statement that resonated the words of the fundamental religious followers who have started criticizing it without understanding the issue, that intrigued Ann Gauger to counter it rationally and she stated "*First of all, who gave scientists the right to interpret Scripture? Why should they care if we believe that we came from a literal first couple? They stuck their noses in where they didn't belong. Second, they didn't test the thing they were claiming.*"

However, the key findings of Hossjer and Gauger (2019) have presented a magnificent narrative of the emergence of the modern human race on Earth, based on genetic data, compatible with a bottleneck from the archaic large population, ended up to a primordial single couple, from which the diverse population of the modern human race has supposedly emerged out, around 500,000 years ago (Sanford and Carter, 2014; Jeanson, 2016; Hossjer and Gauger, 2019). The "Single-couple origin model" has originally been developed and run as a backward simulation model to find a possibility of the emergence of modern human population out of the archaic single couple, either evolved through the trail of bottleneck or emerged from a single first pair (Hossjer and Gauger, 2019). Furthermore, the scientists' duo has done fine trimming of their simulation run, the parsimonious model assumptions of a "Single-couple origin model" and they revised the previous result to claim that the modern human species have supposedly emerged out of the primordial single couple around 100,000 years ago (Hossjer and Gauger, 2019). To examine the Single Couple Origin hypothesis, Hossjer and Gauger (2019) have adopted two simulation models, known as BTSC or Bottleneck to Single Couple and SCO or Single Couple Origin model to run a backward simulation to trigger high pace algorithm, to accommodate larger populations and a large number of variable to yield a feasible and precise result. In their experimental run of the simulation, the scientists have considered the human population as homogenous rather than recognizing the geographic sub-populations, along with recognized other key features like dioecious (like two distinct forms of gender identity, males and females have been recognized) and diploid (two sets of chromosomes in each individual of the population) along with a set of distinctive male (Y) or female (X) chromosome in their genome (Hossjer and Gauger, 2019). Recently, the collation of genetic sequenced data samples (two haploid genome per person sampled) of the genomes of the widespread human population around the world has been done through successful coordination and execution of the projects like International Hap Map Consortium and 1000 Genomes Project Consortium (International Hap Map Consortium, 2007; 1000 Genome Project, 2015) in the field of molecular phylogenetics and human genomics. It needs to be mentioned here the big pool of data, considered here is comprised of millions of genetic variants (multiplied by 5008 haploid genomes in which each variant has been recognized as present or absent here) (Hossjer and Gauger, 2019). In their contemporary investigations, the scientists have preferentially used the published data from 1000 Genomes project (2015) as it has convenient to present the graphical depiction of AFS or Allele Frequency Spectrum as AFS presents the existence of some variants at each proportion (Crow and Kimura, 1970; Ewens, 2004; Durrett, 2008).

Hossjer and Gauger (2019) have studied a comparative account of AFS of African, East Asian and European super populations not only to get an idea on their allele frequencies but getting an overall idea of genetic diversity of the concerned populations and the key observations are mentioned herewith:

a. The variants are mostly rare types and retained by the concerned populations in a small level of frequency.

b. Three super populations have an identical number of common alleles abut a disproportional number of rare alleles.

c. The genetic variation of the African super population is highest among all.

The scientists have engaged in determining the amount of recombination as it has been occurred since emergence from the common ancestor by analyzing the linkage disequilibrium statistics (Hossjer and Gauger, 2019). In the field of cytogenetics, it has been observed that linkage occurs as much genetic information stored nearby and a few numbers of gene variants have found to be copied together. It has been noticed during reduction cell division or meiotic division, that happens between diploid organisms when two sets of chromosomes brought together and crossing over have taken place, yield gametes, possessing "slightly shuffled chromosomes" (Hossjer and Gauger, 2019). It has also been observed that after several generations such as "shuffling" it usually breaks up the linkage between genetic variants and eventually becomes "uncorrelated," recognized as linkage equilibrium (Hossjer and Gauger, 2019). Hossjer and Gauger, (2019) have further contended that "*The linkage or correlation, which is named linkage disequilibrium or LD, fall off with distance between genetic variants, and typically falls off faster distance for an older population*" and that is considered to be an empirical determination to judge whether a population is archaic or derived in nature in terms of the evolutionary scale.

On 28th October 2019, in an exclusive interview with Justin Taylor and Ann Gauger have clarified their objective of coordinating the experimental investigation on backward simulation on BTSC and SCO model and they have referred genetic data is like unique fingerprints, that each individual possesses a unique DNA in its genome, that rendered it to identify and its relatedness to any individual based on identical genetic resemblance. In their interview, scientists have defined human history as a genealogical tree, connecting all contemporary extant humans, where the primordial couple forms the roots of the tree, and the contemporary individuals form the tips of the crown (McVean, 2002). It has also been explained by the investigators that the genealogical tree is not misunderstood to be pedigree as pedigree shows the parental relationships whereas genealogical trees show how inheritance of DNA has likely to be taken place (Taylor, 28th October 2019). To understand the functionalities of inheritance of DNA, Hossjer and Gauger, (2019) have referred as the genetic tracker to go backward to track down the genetic inheritance (identified by a unique mutant tracer) found among the contemporary derived individual (C), derived from its immediate ancestor (B) through vertical inheritance and the immediate ancestor (B), who has received that genetic inheritance from its ancestor (A) like that in a backward cascading manner and the researcher claimed that their computational model has successfully executed it for the entire genome to trace back the history of humanity to go back to million years. Regarding the occurrence of mutations, it needs to be clarified here that it has been started with a single copy-errors during DNA replication but the extent of mutants increased in the population by random occurrence of mutations. Before going further it requires to be introduced with the "HAPLO model," postulated by Hossjer and Gauger (2019). In the HAPLO model, the scientists have tried to define the principles of ancestry in a given population, where each of the population genetically inherits its identity from a single parent (having no recombination) so, it has further been contemplated that a haploid population might have given rise a female population of mitochondrial DNA or a male population of Y-chromosomal DNA. Furthermore, the identical principles of ancestry and mutation have been applied by the scientists in their HAPLO model, for a single locus of DNA, which has been referred as haplotype blocks, in a dioecious, diploid human population, carrying two haploid copies of DNA, inherited each one from its both parents (Hossjer and Gauger, 2019). It needs to be mentioned here that a small-sized recombinant DNA of non-sex chromosomes or the X-chromosomes have been genetically recognized as "Haplotype blocks." As this part of DNA has not gone through recombination, it is considered to be as single locus with a consistent ancestral inheritance of a haploid single population.

Ann Gauger has recognized the HAPLO model as one of the simple, flawless statistical model, brilliantly conceptualized and devised by Ola Hossjer as Gauger has simply defined the functionalities of it, when she has given an interview to Justin Taylor on 28th October 2019, as she (Ann Gauger), stated "*He [Ola Hossjer] would start first from current data and use a method called standard coalescence to trace the lineages backward in time, like growing a forest stems to branches to trunks. Then once he had that forest of trunks, he pruned the picture of all non-productive branches, those that did not have any progeny in the future, the 'dead ends,' and used it as a framework for a new forward-looking model. The pruning saved considerable memory and reduced the computational load. This mean we could go dipper in time or handle larger populations.*" To elucidate the functionalities of HAPLO model, a simple diagram (Fig. 1A and 1B) has been presented here, where the genealogical principle of a haploid population indicates that each individual of the concerned population possess a segment of the chromosome or a single copy of a specific gene and it has also been contended that occurrence of mutations and its accumulations in population-level happens in a constant rate and inherited to generations to generations by descendants. The feasibility of the result, as well as precise determination of time of evolution of any derived group of biological organism, depends on the precise backtracking of the lineages of the most recent generation to its immediate previous generation, from the immediately previous generation to its previous generation to continue the relayed backtracking that all lineages coalesce in one point, that could be considered as the point of origin of the evolutionary journey of the "common ancestor" (yielding the gross time elapsed in between common ancestor on top and the last descendant at the bottom, represent the tentative span of evolution to complete its cycle), of the recently emerged descendants that began its journey from its archaic common ancestor. The population geneticists have claimed that the number of lineages split-out of the common ancestor depends on the size of the ancestral population (Hossjer and Gauger, 2019).

So, this HAPLO model generally defines that the extant individuals of any haploid population, have been emerged from any single ancestor in historical time but it is also found to be true for any consistently large-sized population. So, it rendered the scientists to claim "*...in the latter case we can trace the ancestry of the individuals of the present generation as far back in time as we wish, and sooner or later, with a very high probability, their ancestral lineages will merge at the so-called most recent common ancestor. Therefore we cannot always distinguish a constant population size scenario, from a single individual origin scenario, if their ancestral trees are similar*" (Hossjer and Gauger, 2019). There are some mutational events noticed by the scientists that considered to be the integral features, regulated the population dynamics to some extent (Hossjer and Gauger, 2019):

1. The rate of extinction of mutation is found to be considerably higher in the shrinking population than the expanding population.
2. Most of the newly formed mutation immediately (without being descended to next generations) go extinct rather than the ancestral mutations, normally descended from the historical ancestors to the contemporary descendants.
3. The ancient mutations, happened in historical time, found in fewer numbers but found to be proportionately high in the final generation, whereas the ancestral mutations, recently formed, apparently appeared in big numbers but found to be proportionately few in the final generation.
4. The ancestral mutations that occurred before the existence of common ancestor, seems to be found in the final generation of each individual of a population and got fixed in the population, rather than existing as variants and the fixation of mutation (Kimura, 1987; Lynch, 2007). It rendered the evolutionists to contemplate that "*majority of permanent sequence evolution comes from the random fixation of mutations by drift rather than fixation due to selection*" (Hossjer and Gauger, 2019).

The population geneticists claimed that when a population starts diverging out of a small-sized ancestral population (like the population size of two), it has been recognized as a bottleneck and the occurrence of bottleneck has found to restrict the number of lineages, split-out of the ancestral population. In terms of population genetics, fewer lines in a population mean a lesser extent of its genetic diversity, which leads to bifold genetic casualties like the increasing level of a double recessive and a higher level of inbreeding depression (Hossjer and Gauger, 2019). But the important issue is that it is not the size of the bottleneck that rendered tumble of the genetic diversity of the population of the concerned species but the span of bottleneck that the population sustained, which has experimentally been proved by the scientists and these things happen mostly due to increasing number of harmful variants at the time of bottleneck (Hossjer and Gauger, 2019). After simulating the BSTC model, scientists have noticed that an acute level of the bottleneck of two has brought down the heterozygosity by 25%, as the population would have been bounced back in a short period immediately after bottleneck (Nei et al, 1975). Eventually, a large number of potentially deleterious recessive mutations have also become extinct during this bottleneck but that does not necessarily push the derived species in the brink of extinction. Most often the population geneticists referred the example of a single pair of mouflon that have successfully overcome the impediment of the extreme level of heterozygosity to colonize the entire island, which has found to be a new habitat to it (Keuffer et al, 2007). So, it appears to be a convincible explanation when population geneticists have contended that an extreme bottleneck of two primordial couple, from an erstwhile population of hominids (like chimpanzees), that evolved earlier, would not necessarily end up as an extinct species unless it sustained a long span of the bottleneck (Hossjer and Gauger, 2019).

Hossjer and Gauger (2019) have chosen a tentative model of human origins is a couple, likely to be construed with homozygous genomes, having no primordial diversity at the beginning and gradually accumulate mutations over time and it has been graphically elaborated in three columns as "*Time-evolution of folded allele frequency spectrum: mutations vs primordial variants*." The primary objective of this study is found to show the time-evolution of the folded allele frequency spectra or AFS, which has been compared in three different scenarios (represented by three graphical columns) and it projects the number of density of alleles vs. minor allele frequency. However, out of the three vertical columns, the left one presents the formation of a folded allele frequency spectrum that happens as a result of the accumulation of mutations over time. It is evident that all mutation variants begin as a single copy at the left corner of the population graph and most of the mutations become extinct. Eventually, a counteracting random drift happens in the right corner, got established in the concerned population, whilst the accumulation of new mutations happens on the left side of the population graph. At the point, the researchers noticed that AFS reached an equilibrium state as the number of extinct variants or fix in the population matches the number of new mutations (Hossjer and Gauger, 2019). For this experimental simulation, the size of the effective population size is considered by the investigators around 10,000 and the rate of mutation 48 / haploid genome/generation. The middle and the right columns of this composite population graph present the distribution of existing variation changes over time and between these two columns, the middle column depicts the change of frequency of mutations over time, triggered by genetic drift in a population of 10,000, whilst the right column presents the change of primordial diversity in a population that happens over time in a growing population (growing sigmoidally) from 4,000 to 10,000 in course of a given number of generations, that shows the different growth rate in each graph (Hossjer and Gauger, 2019). In the contemporary investigation, entitled "The genetic effects of the population bottleneck associated with Genesis Flood," the researchers have observed that genetic diversity has been started at 0.5 instead of 0.25 (Carter and Powell, 2016). As the investigators critically watched the genetic drift occurs at a higher pace in small populations, so it has "inverse dependence" in the growth of populations and it has also been observed that the primordial diversity has been looking identical to ancient mutational diversity, as the AFS of existing variant evolves to a flat distribution at 0.5 and 0.25 (Hossjer and Gauger, 2019). The scientists have also observed that when the substantial genetic drift happens in a small population, the distribution of primordially diverse appears to be identical to the distribution of very ancient mutations, yet to be fixed; whereas for large populations it takes long span for ancient mutations to get the state of equilibrium distribution. However, the flat distribution of primordial diversity has rendered the scientists to contemplate that the state of homogenous is homogenous and it grows uniformly, though they have tendered two alternative scenarios of different distributional patterns (Hossjer and Gauger, 2019):

i. The main population might have split into many subpopulations, those have heterogeneous growth patterns and subjected to diverse waves of migration across the continents or

The erstwhile ancestral population might have followed a single wave of colonization, left most of the variations behind.

ii. When the simulation is run for a non-neutral model with selective sweeps with directional selection, each copy of the allele pair of some SNPs might have drifted to fixation that supposedly rendered a part of AFS (originated from primordial variation) resulting ancient mutations to reach the flat equilibrium distribution.

During studying a comparative account of the frequency of the variant in a population of three global super-populations (viz. African, East Asian and European),

Hossjer and Gauger (2019) have observed that based on assumption that the distribution of primordial variants would have been almost flat, the visual estimates of which genetic variants might be primordial in the three global super populations. Scientists have observed a great deal of variation in African DNA than the rest of the two population but none of these three have shown any high-level level of primordially diverse variants or oldest variants in its gene-pool. Furthermore, they have observed that Africans have a higher level of heterozygosity, in comparison to its counterparts so from an evolutionary perspective, these scientists have contemplated the analyzing of AFS and LD of African DNA might help them to get in touch with "more ancient past" of humanity as well as the evolutionary history of humanity on Earth. Hence, they have presented the SCO or Single Couple Origin Model with or without primordial diversity of two diploid parsimonious scenarios in their simulation, that fit in African DNA data, in the perspective of AFS and LD, which is as follows (Hossjer et al 2016a; Hossjer et al, 2016b):

First Scenario: About 2 million years ago, (which is around 100,000 generations ago), a single homozygous couple (i.e. no heterozygosity), grows to a surge of the population of 10,000, followed by a slowdown of growth to reach a 16,000 level at present.

Second Scenario: About 500,000 years ago, (which is around 25,000 generations ago), a single couple have primordial heterozygosity of 0.012 (where the primordial heterozygosity has supposed to be lost due to slow growth rate and the compensation of low primordial diversity would have been done by a surge of growth), grows fast to a size of the population of 16,000 people at present.

The parameters of both scenarios have been chosen based on their fits to the AFS for running HAPLO and they are fit in the LD plot as well. According to the review of a simulation run, in both scenarios, the ancient couple reaches to the population of 4 in first-generation and get doubles at the end of every 10 generations to reach a "specified plateau" and in the real-life human populations of the global population reaches to 8 billion now (Hossjer and Gauger, 2019). However, Hossjer, Gauger, and Reeves have run two large simulations that credibly simulate an entire genome of the human population of 16,000 people for 100,000 generations (Hossjer et al 2016a; Hossjer et al, 2016b). The careful review of the simulation results (briefly narrated in form of AFS and LD analysis) rendered the scientists to claim that human genetic data from non-sex chromosomes are found to be consistently fitted with two distinct parsimonious scenarios of the origin of humanity from a primordial single couple. According to the first scenario, the first couple without diversity would have possibly lived about 2 million years ago and according to the second scenario, the single couple with primordial diversity would have supposedly lived about 500,000 years ago. So, the backward simulations, run in the HAPLO algorithm by a group of evolutionary scientists like Hossjer, Gauger, and Reeves might be considered as a minority group of scientists who have successfully tested the "Single-couple origin" hypothesis and found to be successful in upholding it as a feasible possibility, despite sustaining the strong objections and the barrage of negative comments from the "science over religion" group of evolutionists (actually the majority of the evolutionary biologists belong to this category) (Hossjer et al 2016a; Hossjer et al, 2016b; Hossjer and Gauger, 2019). This conclusion of this simulation is intriguing as the SCO model is found to be a simple parsimonious simulation model (the possible extension of designed diversity model) which has successfully match the data to the assumptions and the most important observation of this empirical investigation successfully challenged the erstwhile assumptions that the modern human species had exclusively evolved out of non-human primates (like Great Apes) of big populations, rather than it is more convincible that emergence of modern human species descended out of archaic hominins, (where substantial evidence of interbreeding is there in between Neanderthals and Denisovans) in historical time and the genetic diversity in the derived or contemporary population of modern human species come to the descendants from ancestors by cascading cycles of germline mutation, resulting in vertical inheritance (Sankararaman et al, 2014; Yang et al, 2014; Danneman et al, 2016; Dechamps et al, 2016). In his exclusive interview with Justin Taylor, Ola Hossjer has made a convincing state, as he firmly stated "*It is hard to prove that a certain model of human history is the right one based on genetic data from individuals of today. The reason is that sometimes quite different models of human ancestry might fit genetic data equally well. We showed that a model with a first unique couple gave a good fit to some African genetic data. Therefore we cannot rule out a model where humanity started from the first couple in favor of a model where we share ancestry with chimps and other species*" (Taylor, October 28, 2019).

References

Ayala, F. J. (1995). The myth of Eve: molecular biology and human origins. *Science.* 270(5244): 1930-6. doi: 10.1126/science.270.5244.1930.

Ayala, F. J., and Escalante, A. A. (1996). The evolution of human populations: a molecular perspective. *Mol Phylogenet Evol.* 5(1):188-201. doi: 10.1006/mpev.1996.0013.

Ayala, F.J., Escanlante, A., O'hUigin, C. and Klein, J. (1994). Molecular genetics of speciation and human origins. *PNAS USA* 91: 6787-6794.

Barras, C. (March 6, 2013). The father of all men is 340,000 years old. https://www.newscientist.com/article/dn23240-the-father-of-all-men-is-340000-years-old/

Bergström, T., Josefsson, A., Erlich, H., and Gyllensten, U. (1998). Recent origin of *HLA-DRB1* alleles and implications for human evolution. *Nat Genet* 18: 237–242. https://doi.org/10.1038/ng0398-237

Bersaglieri, T., Sabeti, P.C., Patterson, N., Vanderploeg, T., Schaffner, S.F., Drake, J.A., Rhodes, M., Reich, D.E., and Hirschorn, J.N. (2004). Genetic signatures of strong recent positive selection at the lactase gene. *Am. J. Hum Genet.* 74(6): 1111-20.

Blum, M.G.B., and Jakobsson, M. (2011). Deep divergences of human gene trees and models of human origins. *Mol Biol Evol* 28: 889–898.

Carter, R.W., Lee, S.S. and Sanford, J.C. (2018). An overview of the independent histories of the human Y-chromosome and the human mitochondrial chromosome. In *Proceedings of the Eighth International Conference on Creationism*, ed. J.H. Whitmore, pp. 133–151. Pittsburgh, Pennsylvania: Creation Science Fellowship.

Carter, R.W. and Powell, M., (2016).The genetic effects of the population bottleneck associated with the Genesis Flood, *J. Creation* **30**(2):102–111.

Coelho M., Luiselli, D., Bertorelle, G., Lopes, A.I., Sexias, S., Destro-Bisol, G., and Rocha, J. (2005). Microsatellite variation and evolution of human lactase persistence. *Hum. Genet.* 117(4): 329-39.

Crow, J. F., and M. Kimura, (1970). *An Introduction to Population Genetics Theory.* Harper and Row, New York.

Dannemann, M., Andrés, A.M., and Kelso, J. (2016). Introgression of Neandertal- and Denisovan-like Haplotypes Contributes to Adaptive Variation in Human Toll-like Receptors. Am J Hum Genet. 98(1):22-33. doi: 10.1016/j.ajhg.2015.11.015.

de Groot, R., Brander, L., van der Ploeg, S., Costanza, R., Bernard, F., Braat, L., Christie, M., Crossman, N., Ghermandi, A., Hein, L., Hussain, S., Kumar, P., McVittie, A., Portela, R., Rodriguez, L.C., ten Brink, P., and van Beukering, P. (2012). Global estimates of the value of ecosystems and their services in monetary units. Ecosystem Services. 1(1): 50-61,ISSN 2212-0416, https://doi.org/10.1016/j.ecoser.2012.07.005.

Deschamps, M., Laval, G., Fagny, M., Itan, Y., Abel, L., Casanova, J. L., Patin, E., and Quintana-Murci, L. (2016). Genomic signatures of selective pressures and introgression from archaic hominins at human innate immunity genes. *Am J Hum Genet.* 98:5–21.

Dorak, M. T., Lawson, T., Machulla, H. K., Mills, K. I., and Burnett, A. K.(2002). Increased heterozygosity for MHC class II lineages in newborn males. *Genes Immun.* 3(5): 263-9. doi: 10.1038/sj.gene.6363862.

Doxiadis, G. G., Hoof, I., de Groot, N., Bontrop, R. E. (2012). Evolution of HLA- DRB genes. *Mol Biol Evol.* 29(12):3843-53. doi: 10.1093/molbev/mss186.

Durrett R. (2008). *Probability models for DNA sequence evolution.* New York: Springer.

Enns, P. (2012). *The evolution of Adam: What the Bible does and doesn't say about human origins.* Baker books. P. 192.

Ewens, W. J., (2004). Mathematical population genetics, in Interdisciplinary Applied Mathematics, Vol. 27. Springer, New York.

Gauger, A., Axe, D., and Luskin, C. (2012). Science and Human Origins. Discovery Institute Press (Seattle). P.126.

Gerbault, P., Liebert, A. Itan, Y., Powell, A., Currat, M., Burger, J., Swallow, D.M., and Thomas, M.G. (2011). Evolution of lactase persistence: an example of human niche construction. *Philos trans R. Soc. Lond. B. Biol. Sci.* 366(1566): 863-77.

Hawks, J. (March 3, 2020). Quote: No 'anatomically modern' elephants. https://johnhawks.net/weblog/topics/history/history-of-paleoanthropology/tobias-anatomically-modern-elephants-2020.html

Hawks, J., Wang, E.T., Cochran, G. M., Harpending, H.C. and Moyzis, R. K. (2007). Recent acceleration of human adaptive evolution. *PNAS USA* 104(52): 20753-20758.

Hössjer, O. and Gauger, A. (2019). A single-couple human origin is possible. *BIO-Complexity.org* 2019(1): 1-20.

Hössjer, O., Gauger, A., and Reeves, C. (2016a). Genetic modeling of human history part 1: comparison of common descent and unique origin approaches. BIO-Complexity 2016 (3):1–15. doi:10.5048/BIO-C.2016.3.

Hössjer, O., Gauger, A., and Reeves, C. (2016b). Genetic modeling of human history Part 2: A unique origin algorithm. *BIO-Complex-ity* 2016 (4):1–36. doi:10.5048/BIO-C.2016.4

International Hap Map Consortium, (2007). A second generation human haplotype map of over 3.1 million SNPs. *Nature* 449: 851-861.

Kueffer, C., Hadorn, G. H., Bammer, G., van Kerkhoff, L., and Pohl, C. (2007). Towards a Publication Culture in Transdisciplinary Research. *GAIA -*

Ecological Perspectives on Science and Society 16(1). DOI:10.14512/gaia.16.1.8

Itan, Y., Powell, A., Beaumont, M.A., Burger, J., and Thomas, M.G. (2009). The origins of lactose persistence in Europe. *PLoS Comput Biol.* 5(8): e1000491.

Jeanson, N.T. (2016). On the Origin of Human Mitochondrial DNA Differences, New Generation Time Data Both Suggest a Unified Young-Earth Creation Model and Challenge the Evolutionary Out-of-Africa Model. https://www.semanticscholar.org/paper/On-the-Origin-of-Human-Mitochondrial-DNA-New-Time-a-Jeanson/3857a6e05f2589065c7d3a97bd7244028befea30

Kaimen-Maciel, D. R., Reiche, E. M., Borelli, S. D., Morimoto, H.K., Melo, F.C., Lopes, J., Dorigon, R.F., Cavalet, C., Yamaguchi, E.M., Silveira, T.L., Da Silva, W.V., Comini-Frota, E.R., Brum Souza, D.G., Donadi, E.A. (2009). HLA-DRB1* allele-associated genetic susceptibility and protection against multiple sclerosis in Brazilian patients. *Mol Med Rep.* 2(6): 993-8. doi: 10.3892/mmr_00000204.

Kimura, M. (1969). The number of heterozygous nucleotide sites maintained in a finite population due to a steady flux of mutations. *Genetics* 61: 893-903.

Kimura, M. (1987). Molecular evolutionary clock and the neutral theory. *J Mol Evol* 26: 24–33 https://doi.org/10.1007/BF02111279

Kimura, M. and Ohta, T. (1971). Protein polymorphism as a phase of molecular evolution. *Nature* 229(5285):467-469.

Lynch, M. (2007). The frailty of adaptive hypotheses for the origins of organismal complexity. *PNAS USA* 104(1): 8597- 8604.

McVean, G. A. (2002). A genealogical interpretation of linkage disequilibrium. *Genetics.* 162(2): 987-91.

Nachman, M. W., and Crowell, S. L. (2000). Estimate of the mutation rate per nucleotide in humans. *Genetics* 156(1):297-304.

Nei, M., Maruyama, T. and Chakraborty, R. (1975). The Bottleneck effect and genetic variability in populations. *Evolution* 29(1): 1-10.

Rafajlović, M., Klassmann, A., Eriksson, A., Wiehe, T., and Mehlig, B.(2014). Demography-adjusted tests of neutrality based on genome-wide SNP data. *Theor Popul Biol.* 95:1-12. doi: 10.1016/j.tpb.2014.05.002.

Sanford, J. C. and Carter, R. (2014). In light of genetics…Adam, Eve, and the Creation/fall. *Christian Apologetics Journal* 12 (2): 51-72.

Sankararaman, S., Mallick, S., Dannemann, M., Prüfer, K., Kelso, J., Pääbo, S., Patterson, N., and Reich, D. (2014). The genomic landscape of Neanderthal ancestry in present-day humans. *Nature.* 507(7492):354-7. doi: 10.1038/nature12961.

Taylor, J. (October 28, 2019). Does Science Disprove Adam and Eve? An Interview with a Biologist and a Mathematician.

https://www. thegospelcoalition.org/blogs/justin-taylor/new-scientific-study-showing- come-original-adam-eve-interview-authors/

The 1000 Genomes Project Consortium. (2015) A global reference for human genetic variation. *Nature* 526: 68–74. https://doi.org/10.1038/nature15393

Titus-Trachtenberg, E.A., Rickards, O., De Stefano, G.F., and Erlich, H.A. (1994). Analysis of HLA class II haplotypes in the Cayapa Indians of Ecuador: a novel DRB1 allele reveals evidence for convergent evolution and balancing selection at position 86. *Am J Hum Genet*. 55(1):160-7.

Venema, D.R. and McKnight, S. (2017). *Adam and the genome: Reading scripture after genetic science.* Brazor press. P. 221.

von Salomé, J., Gyllensten, U. and Bergström, T.F. (2007). Full-length sequence analysis of the HLA-DRB1 locus suggests a recent origin of alleles. *Immunogenetics* 59: 261–271.https://doi.org/10.1007/s00251-007- 0196-8

Vuorisalo, T., Arjamaa, O., Vasemägi, A., Taavitsainen, J. P., Tourunen, A., and Saloniemi, I. (2012). High lactose tolerance in North Europeans: a result of migration, not in situ milk consumption. *Perspect Biol Med.* 2012; 55(2): 163-74. doi: 10.1353/pbm.2012.0016.

Wilson Sayres, M. A., Lohmueller, K. E., and Nielsen, R. (2014). Natural selection reduced diversity on human y chromosomes. *PLoS Genet.* 10(1):e1004064. doi: 10.1371/journal.pgen.1004064.

Wright S. (1938). Size of population and breeding structure in relation to evolution. *Science* 87: 430–431.

Yang, M.A., Harris, K., and Slatkin, M. (2014). The projection of a test genome onto a reference population and applications to humans and archaic hominins. *Genetics* 198:1655–1670.

Yuan, Q., Chu, C., and Jia, J. (2012). Association studies of 19 candidate SNPs with sporadic Alzheimer's disease in the North Chinese Han population. *Neurol Sci.* 33(5):1021-8. doi: 10.1007/s10072-011-0881-0.

Chapter 11

Delving for Metaphoric Garden of Eden: Humanity's First Homeland on Planet Earth - I

It is obvious that modern humans did not interbreed with hominins living over 500,000 years ago. It is also clear that there was no single Adam and Eve but rather groups of Adams and Eves living side by side and wandering together in our world.

---- Eran Elhaik

Let us start with the mutual antagonistic role-play of biology and culture to recognize any possibility for better understanding whether the origin and evolution of human civilization on planet Earth should be considered as natural or unnatural. Biology plays a permissive role when conceiving the existence of some possibilities. Whereas the cultural bottlenecks have let some of these possibilities to be designated as natural options but forbid other possibilities to be designated as unnatural possibilities. Whether it's socio-culturally recognized as married or unmarried, biology has enabled all matured women to have children, so this specific biological possibility should be considered as natural traditions. Some socio-cultural and/or socio-cultural practices have recognized it as natural to conceive a baby if the women are married but considered unnatural customs when they are not married. Biologically, humankind (*Homo sapiens*) has broadly been distinguished as males (having one X, and one Y chromosomes) and females (having two X chromosomes). Biology enables the possibility of having sex among matured males and females, and biology has never restricted the sexual interactions between two male or two female individuals. Furthermore, biology has enabled the possibility that in the normal course of sexual interactions between two individuals of identical sexual profile (whether it is between two males or two females) as well as an act of homosexuality would not help to reproduce any future progeny, but an act of heterosexual (male-female interaction) could result in the reproduction of a new offspring. So, biology has come up with possibilities of homosexual or heterosexual sexual interaction among male and female individuals belonging to Homo sapiens as natural course of actions but some socio-religious and socio-cultural practices forbid the act of homosexuality as forbidden and unnatural, whereas they accept and promote the act of heterosexuality as natural course of action in humankind. So, it's clear that the consideration of any possibility to recognize whether it natural or unnatural has never been guided by biology, rather it is determined by the theological, socio-religious practices and socio-cultural norms of diversified communities, sects, clans of humankind, dispersed all over the world.

So, it seems that it would be a straight-forward explanation to study the typical chromosome (genotypic) profile of individual Sapiens and recognize the concerned individual as a male or female member of Sapiens. Rather it would be considered as the "lock and key" compatibility model where the characteristic genotypic profile, as well as a biological trait, was found to be compatible with the cultural myth to recognize the concerned Sapiens as a male or female representative. Suppose, a typical biological trait of a Sapiens individual with a set of XY chromosomes, the preponderance of testosterone hormone, mustache and a muscular body fits into the cultural (or socio-cultural) myth of masculinity, like a fighting spree, engaging in trans-border trading, and participating in active politics, etc. has helped it to fit in the slot of male representatives of Sapiens according to the human society's well-defined order. Likewise, a Sapiens individual with the characteristic biological traits of XX genotypic profile, with characteristic secondary reproductive organs (like the presence of breast in upper part and womb in the lower part of the body), the preponderance of estrogen hormone, non-muscular, and a body with a heavy waste-line fits into the cultural (socio-cultural) myth of feminine features like staying home, obeying the male partners, giving birth to and rearing the children, etc. have helped to fit it in the slot of female representative of Sapiens, according to the human society's well-defined order. So, the recognition of manhood or womanhood as maleness and femaleness is an instance of socio-cultural recognition, that varies from one society to another one and changes over time and it helps to understand that "sex" is considered to be a biological division, whereas "gender" should be recognized as a cultural category.

Let us try to realize any possibilities in terms of natural or unnatural perspective. Most of the theologists argued that the creation of the human body has been closely integrated with the creation of specific limbs for purposeful execution of any specific task and it happened as intended by the Holy Spirit of God. Hence the typical use or functioning of any specific limb or part of the human body is supposedly considered as natural, as long as it complies with the prototypic pattern, considered to be the ultimate using pattern intended by God. So, an effort of the alternative use of any limb, part of the body or any particular organ by a human being by stepping out of its prototypic or archaeo-typic usage pattern would be considered as unnatural as it might not comply with God's intension. From an evolutionary perspective, the origin and evolution of any part or organ of the human body has been considered as a random process that has no distinct purpose. Furthermore, the evolutionists have argued that the origin and evolution of any organ in any organism has usually attributed with a prototypic task when it emerged in any archaic, living beings around hundreds of thousands of years ago, followed by several behavioral and functional changes in course of adaptive changes to cope and sustain in the survival race of evolution. The evolutionary biologists have tendered an example of the evolution of mouth in the archaic annelids or pre-historic worm-like animals, mainly for execution of the prototypic function of food-intake (nutrients) to survive; whereas the presence of a mouth in the human species has supposedly gone through many evolutionary diversifications and bottlenecks of adaptations to functionally appear as the key organ for verbal communication by emitting sound waves and expressing love and affection by using it for kissing and smooching. So, over time, any organ or limb, which evolved with a typical prototypic function, has gone through the evolutionary labyrinth of diversifications and emerged to be with key physiological and behavioral functionaries for executing the unique type of function, and it does not help to put any identifying archaic or prototypic function (performed by any specific organ or set of organs) as natural and the derived or recently evolved functions as unnatural. So, from an evolutionary perspective, any effort of moral judgment or putting an identifying tag as natural or unnatural on "Sex" is an ambiguous state of perception. In his provocative, popular nonfiction, "*Sapiens: A Brief History of Humankind*," Yuval Noah Harari (2011) has logically argued "*Sex first evolved for procreation and courtship rituals as a way of sizing up the fitness of a potential mate. But many animals now put both to use for a multitude of social purposes that have little to do with creating little copies of themselves. Chimpanzees, for example, use sex to cement political alliances, establish intimacy and defuse tensions, Is that unnatural?.*"

Hence, if we consider two distinct kinds of sexual orientation in the community of Sapiens, either of heterosexual orientation between male and female individuals or of homosexual orientation between two male or two female individuals, it would be recognized as two biological possibilities of sexual orientation; but in terms of socio-cultural or socio-religious perspective, when heterosexual orientation has been considered as natural, homosexuality has been stigmatized as unnatural in terms of human social order. Though male (having XY chromosomal profile and distinctive structural orientation) and female (having 2X chromosomal profile and distinctive structural orientation) representative of Sapiens have a common and archaic history, origin and evolutionary history, men and women have gone through distinctive survival and reproductive orientation in its trail of evolution and diversification, it appeared to a consecutive representative of two distinct genders in the community of Sapiens. Though, exception is always there, the evolutionary history of Sapiens for hundreds of thousand years, hardly portrays women as an aggressive representative of Sapiens which involved competing with other women to fight and outperform other female competitors, and to pick its male counterpart as the most eligible, fertile, sexual partner, rather the conventional sexual orientation of men has portrayed it as the most aggressive representative in Sapiens, those involved in competing against other male competitors to win over the most competent, fertile, female partner for sex and the phenotypic expression of such ambitious, pro-aggressive (though it has been changed a lot in men through passage of time) evolutionary socio-behavioral traits have supposedly been inherited and carried forward in the "masculine genes" of the genome in the male archaic predecessors of the contemporary descendants, and transcended from the prehistoric predecessors to the contemporary successors for generation after generation (Harari, 2011). Whereas the evolutionary history of Sapiens, particularly the feminine members of Sapiens, would focus on the evolutionary process of diversification of "feminine genes," possessed in the genome of women, have supposedly rendered them more submissive in respect to their male counterparts, mostly due to dealing with the issues like possessing an embryo in their wombs, and giving birth, that made them confine inside their dwelling place for a long period (around 9 months). Furthermore, such a long period of staying home and less mobility would render the female representative of Sapiens bio-behaviorally less aggressive than their male counterparts and dependent upon their sex partners for ensuring food and safety. When the possessor of masculine genes has more access to step out of and explore uncharted territory and participated ambitiously to take part in course of migration, the possessor of "feminine genes" would have played a passive role in that. Hence, it is convincible that the ambitious and aggressive metaphoric expansion mode of "masculine genes" has played a cardinal role in the origin and evolutionary diversification of Sapiens (specifically emergence of new races of cross-cultural identity across the globe) on this blue planet since its origin in comparison to its not too-aggressive "feminine genes."

Let us go back to square one of molecular biology to conceive the perception of "Gene," metaphorically referred to as the story-book, which has been written in the language of "Nucleic acid," specifically "DNA" (Deoxyribose Nucleic Acid). From an evolutionary perspective, the gene has been recognized as the ultimate structural and functional unit of heredity. Generally speaking, genes are considered to have protein-encoding potential, but scientists have witnessed some genes having no protein encoding potential and the proteins are found to carry out specific functions in one or more types of cells in the living organism. Genes of any individual have been contained by the chromosomes, found in the cell, specifically the cell nucleus. According to the recent findings on Human Genome Project, there are 20,000-25,000 genes in the human genome. Every Sapiens individual has possessed any particular gene in two copies, each one inherited from each parent. Though around 99% of genes in the gene pool of the human population is found to be identical, yet a small part possessed by each individual (which is around 1%) is found to be slightly different, and these little differences in a genetic profile at an individual level, attributes physical uniqueness of each Sapiens. In genetic research, scientists have preferred to refer to one of the most frequently used terminologies, called "Allele," which appeared as a varied form of the same gene, characterized by the minute differences in their sequences in their DNA bases. Since the 1980s, DNA has randomly been used as the most preferred tool by scientists from interdisciplinary areas of molecular genetics, genome biology, evolutionary genetics, and genomics in digging out the history of origin and evolution of the modern human species. The most interesting part of these DNA related investigations, carried out by the scientists, have primarily focussed on analyzing the genes, contained by mitochondria (considered to be the ultimate powerhouse of the cell), and each of these cell organelles, which carries some unique, circular-shaped, ring-like DNA, called mitochondrial DNA or plasmid DNA. Molecular biologists have preferred to carry out their investigations with mitochondrial DNA, rather than undertaking the cumbersome process of extraction and analysis of the DNA sample from the cell nucleus. The interdisciplinary investigators in the field of genome biology, molecular genetics, evolutionary biology, and developmental biology have found to be successful in determining the tentative time of origin and evolution of mitochondrial DNA or extra-nuclear DNA, which had likely been synthesized as well as evolved as a result of accidental but successful introgression (or engulfing) of an oxygen-breathing prokaryotic bacterium by a unicellular eukaryotic organism, to accomplish formation of an endosymbiotic cellular organelle (like Mitochondria and plastid) around 2 billion years ago. Eventually, such archaic, unicellular, eukaryotic (accidentally evolved endosymbiont) entity has been contemplated as the prehistoric ancestor of anatomically modern human representative of Sapiens. It has been witnessed by the reproductive biologists that the successful effort of physical mating between male and female Sapiens has ended up with the initiated fertilization of an egg by a sperm. The male gamete is found to end up with the formation of an embryo or the new progeny, that started growing inside the womb of the female Sapiens. The formation of the embryo is found to be accomplished during fertilization with a successful injection of its nuclear DNA (haploid or n), possessed in the male gamete to the female gamete, followed by fusion of two haploid nuclear DNAs to form embryonic nuclear DNA (of diploid or 2n), but it failed to accomplish transfer and fusion or recombination of mitochondrial DNA during fertilization. Eventually, scientists have noticed that in the post-fertilization stage, the entire mitochondrial elements, possessed in the female gamete or egg, have been remained unaltered (due to lack of transfer of mitochondrial DNA or Mt DNA elements from male to female gamete, and no further chance of fusion of Mt DNA from male and female Sapiens) and it rendered retention of Mt DNA by the female Sapiens that not only contended in the zygote, is conceived, but it has also maternally inherited by the future offsprings.

Hence, the inherited copies of mitochondrial DNA in any offspring is considered in being inherited from its mother and biologically recognized as an instance of maternal inheritance, which occurs outside of the fertilization cycle; whereas the inheritance of nuclear DNA is found to accomplished through an act of fertilization in the embryo of the womb, after being transcended from its paternal side to be infused with the nuclear DNA of its counterpart, the female Sapiens. The intermittent inheritance of mitochondrial DNA, that has carried forward the features of maternal inheritance from mother to offspring has been continued for generations after generations unless it had gone through the bottleneck of mutational changes, which is the normal occurrence, that any biological entity would traverse through the temporal evolutionary trail of few hundred of thousand years (Zimmer, August 23, 2004). As every biological entity would hardly escape the gateway of mutational alterations, mitochondria, the cellular organelle is no exception. Hence, the mutated mitochondrial DNA has intermittently transcended from mother to offspring for many generations, leading the evolutionary diversification of the archaic predecessors to have emerged as the unique descendants of distinct molecular foot-prints (that would inevitably distinguish from the archaic molecular foot-prints of the predecessors). Hence, the meticulous genealogical investigation on such, mutation-driven altered molecular foot-prints of mitochondrial DNA, extracted and analyzed from the offsprings by molecular geneticists and genomicists have helped further to decrypt the evolutionary history of Sapiens, that started with the primordial women, from whom we have inherited the mitochondrial DNA, employing maternal inheritance and continuously going through the changes (Zimmer, August 23, 2004). In this regard, the mutational research studies on the maternal inheritance of Sapiens, carried out by Prof. Allan C. Wilson, Department of Biochemistry, University of Berkley, USA, who has meticulously examined the mtDNA samples, randomly extracted from 147 individuals, dispersed all-over a wide geographical areas of Africa, Asia, Australia, Europe, and New Guinea to re-construct a molecular family tree. Prof Wilson's experimental result is quite intriguing as it indicated that out of 174 randomly picked individuals, an unusual kind of identical mutations has been witnessed in the genomes of four individuals of geographically distinct locations. Hence, it inevitably indicated further that archaic lineages of Sapiens must have been descended from a prehistoric mother-stock. Though an alternative hypothetical explanation might be also be considered, that an unusual kind of mutation which happened in the identical location of the genes of such four distinct individuals, accidentally emerged in different geographical locations at the same point of time; but there are some 'but's and 'if's to consider it as a feasible hypothesis (Zimmer, August 23, 2004).

A unique endeavor of reconstruction of the family tree of Sapiens (from several individuals of far and wide geographical locations), which was undertaken by Douglas C. Wallace, one of the leading evolutionary and molecular geneticist from the University of Pennsylvania and Children's Hospital of Philadelphia in Pennsylvania, USA, and his teammates, demonstrated that out of all major evolutionary lineages of Sapiens have likely been merged to 5 major maternal genetical lineages, and these 5 genetical lineages supposedly emerged out of the respective continents, supposedly coalesced to a pre-historic mother-stock, but the intensive phylogenomics investigation has elucidated it further that there are 7 individuals of Sapiens of African descent have supposedly emerged out of another major evolutionary lineage (Zimmer, August 23, 2004). Wallace and his research colleagues empirically determined that in their evolutionary reconstruction model of the family tree of Sapiens, the African evolutionary lineage has supposedly gone through the mutational a lot more than two times in respect to the evolutionary lineages of Sapiens from Asia and Europe (Zimmer, August 23, 2004). Furthermore, this observation of Wallace and his research colleagues have rendered the phylogeneticists and genomicists to extrapolate that the origin of humanity, as well as the emergence of the archaic lineage of Sapiens, have supposedly emerged out of the African continent, and it has gone through the evolutionary divergence in course of time. It has been reasonably hypothesized that the evolution of several derived lineages (of African descent) of Sapiens out of the main archaic African lineage which spread out of these lineages in the far and wide, congenial, terrestrial ecosystems on this Blue Planet, happened as a result of intercontinental migration and this maiden endeavor of pre-historic migration of Sapiens out of its place of origin and has been referred as "Out of Africa" (Zimmer, August 23, 2004; Groucutt et al, 2015).

Later, some scientists have tendered much evidence in support of the "Out of Africa" hypothesis, about the origin and evolution of predecessors of the anatomically modern representative of Sapiens out of the African continent and some of these are unique. After critically examining and reviewing the anthropological pieces of evidence and studying the prehistoric cave paintings, the paleo-anthropologists have contemplated that the anatomically modern representatives of Sapiens have migrated to the Asian continent (particularly in Indonesia) from the African continent and finally moved to Australia around 50,000 YBP (Meyerhoff, 2013). In this regard, it's worth mentioning one of the wild, strong, and robust archaic predecessor, known as the Neanderthal man (or *Homo neanderthalensis*) of the modern Sapiens, who reportedly endured the most adverse, environmental condition in the Ice Age around 250,000 years ago, their footsteps have been tracked in Europe by the paleo-anthropologists, supposedly migrated from Asia (Meyerhoff, 2013). The most intriguing part of this story, which relates to the so-called extinction or disappearance of this archaic representative of Sapiens, has been shared by the paleoanthropologists and been narrated that around 10.000 YBP, Neanderthal men started disappearing mainly due to the intimate relationship, as well interbreeding between *Homo neanderthalensis* and *Homo sapiens*, that caused the genetic introgression of this archaic human species in the genome of modern representative of Sapiens (Meyerhoff, 2013). Going a step forward, the paleoanthropologists have referred to the burial site of Quafez (adjacent to Nazareth, Israel), one of the oldest burial sites, close to Africa, and the age-dating process determined it around 100,000 years old. Eventually, the paleoanthropologists succeeded to study the grave of the archaic representative of Sapiens, which was found to be 60,000 years old and it helped the researchers to reconcile the tentative maiden endeavor of migration of the archaic representative of the Sapiens, out of Africa (Gugliotta, July 2008). However, the contemporary advancement in the field of molecular genetics, evolutionary genetics, and phylogenomics, particularly in analyzing the DNA sequencing data either from mtDNA or from Y Chromosome of Sapiens has helped in the genealogical reconstruction of the archaic state of the modern representative of Sapiens. The reconstruction of the family tree of Sapiens has normally been undertaken, primarily based on the genetic differences, witnessed in the non-coding regions of DNA. Furthermore, the scientists have witnessed that the genes in the non-coding region of DNA have not only been found to have the least effect on the body of human beings, in terms of molecular genetics, but the determination of evolutionary age of any archaic lineage of

Sapiens are also found to be little chaotic, mainly due to the source of DNA sample analyzed and the distinct mode of calibration of the molecular clocks, standardized by the molecular phylogeneticists. Hence, the determination of the evolutionary age of any archaic lineage of Sapiens is yielding unique results, whenever the scientists have employed different calibration strategies to utilize their molecular clocks. Yet, the empirical values of the evolutionary ages on such genealogical reconstruction of the archaic lineage of Sapiens have not been overlooked. The collective results on several archaic evolutionary lineages of Sapiens on such genealogical investigations have appeared to be as projected backtracking of all evolutionary lineages, coalesced to a prehistoric Spatio-temporal matrix, whenever the archaic lineages of Sapiens likely to split from each other. Based on the analysis of such genealogical reconstruction, the molecular biologists have tried to figure out the tentative prehistoric time, whenever the multiple archaic lineages of Sapiens started branched out from each other and tried to backtrack at the same time to reach out to the geographic location (coalesced to a starting point), recognized to be the erstwhile location of origin and evolution of Sapiens, and experimentally it has been investigated and determined by the different researchers the African continent for the maiden birth-place of Sapiens. So, the experimental observations on some genealogical reconstructions, have indicated further that the origin and evolutionary marathon of Sapiens, which has metaphorically started from "Eden," and the birthplace of modern representative of Sapiens, is found to be identified as somewhere in the African continent. In support of such hypothetical observation like "Universal African Brotherhood" about the origin and evolution of the modern-day Sapiens, the observation of Dr. Peter A. Underhill, an evolutionary biologist from Stanford University, USA, who engaged in genealogical reconstructions of Sapiens for a long time (based on the analysis of DNA sequences of the Y chromosome), that rendered him to make an unequivocal and philosophical statement "*We are all Africans at the Y chromosome level and we are all brothers*" (Wade, May 2, 2000).

So, the earnest endeavor of genealogical reconstruction of the family tree of Sapiens has successfully been accomplished by a group of molecular geneticists and evolutionary biologists, led by Douglas C. Wallace, from University of Pennsylvania and Children's Hospital of Philadelphia in Pennsylvania, USA, who had successfully used the mitochondrial DNA as a molecular marker in the successful completion of their evolutionary investigations (Wade, May 2, 2000). Dr. Wallace and his research colleagues published their research observation on the genealogical reconstruction of Sapiens in the "American Journal of Human Genetics" in March 2000, and they have tried to elucidate the root of the mitochondrial tree, which has found to be occupied by "Kung San," one of the archaic, nomadic, population of Sapiens as well as the African hunter-gatherers of North-western Kalahari desert in South Africa (Meyerhoff, 2013). Furthermore, Wallace and his teammates observed that the population of "Biaka pygmies" of Central African origin, one of the prehistoric representatives of Sapiens, have found to be endemic in the distribution in the identical geographical location in Africa and the evolutionary age has found to be almost identical to the "Kung San" community of Sapiens. Based on their meticulous research reviews, Wallace and his research colleagues have further contemplated that there are no or little changes that have been witnessed in the profiles of mtDNAs of the prehistoric populations of Sapiens in respect to their predecessors where relative isolation of such archaic populations of Sapiens have played no or little impact (Wade, May 2, 2000). These unique research findings on the genealogical reconstruction of family trees of Sapiens (based on the critical analysis of the mtDNA, randomly extracted from the archaic populations of Sapiens) have inspired Dr. Wallace and his team members to understand and narrate eloquently, the introductory part of the history of origin and evolution of anatomically modern human species or Sapiens on Earth and it inspired Dr. Wallace to comment "We are looking at the beginning of what we would call Homo sapiens" (Wade, May 2, 2000).

Professor Bryan Sykes, the eminent molecular phylogeneticists from Oxford University, and his research colleagues have traced out seven prehistoric mother groups in Sapiens or seven matriarchal lineages of Sapiens, supposedly acting as the mother stock of all European descendants, stoically recognized in the science/non-fiction "*Seven European Daughters of Eve*" (Sykes, 2001). Furthermore, Professor Sykes and his co-researchers have focussed on the paleoanthropological identity of these prehistoric, matriarchal lineages and firmly claimed that each of "Seven European Daughters of Eve" emerged out of the "Lara clan," considered to be one of the ancestral clans of Sapiens on Earth, specifically originated and evolved on the African continent (Meyerhoff, 2013). In his popular non-fiction "*The*

Story of Us," Janusz Meyerhoff (2013) elaborated on the research findings on human evolution by Prof. Sykes and his research colleagues, particularly on the arena of mitochondrial genetics of Sapiens, to make it convincible for a general audience to grasp how the contemporary analysis and interpretation of such analysis on the ancient DNA of Sapiens would define the origin of the archaic population of Sapiens and tracing back the evolution of anatomically, modern human species from its predecessors. In their meticulous investigations with mtDNA samples, Professor Sykes and his research colleagues have randomly collected around 6000 mtDNA samples from the buccal cavity of a large community of Sapiens. After critical examination and analyzing the results on DNA sequencing, of these mtDNA samples, Professor Sykes has contemplated that the archaic, matriarchial lineages of the Sapiens existed on Earth between 8,000-45,000 years ago (Meyerhoff, 2013). It is quite fascinating for the molecular phylogeneticists to get positive results in the experiments on the genealogical reconstruction of family trees of Sapiens by analyzing the mutational rate of mtDNA in Sapiens. The main strategic advantage in such experiments, laid in the lesser extent of the rate of mutation, that mtDNA of Sapiens have gone through, are found to be much less (supposed to range from 1-10,000 years), in comparison to the nuclear DNA.

Normally, most of the academic researches have remained untapped to exploit its commercial potential, but the academic researches of Professor Bryan Sykes and his research colleagues are found to be an exception in the field of applied researches as in April 2000, "www. Oxfordancestor.com," a U.K. based business start-up had started its operation with a unique offer to its potential customer to introduce them to their prehistoric predecessor, by establishing the evolutionary relationship to a specific matriarchal lineage of seven daughters of Eve, they have descended from. These unique business ideas of Professor Bryan Sykes of the University of Oxford had made him the founder as well as the successful entrepreneur in an uncharted territory of applied science like Genealogical investigating start-up, as Sykes has the strategic advantages to deal with the mitochondrial lineages of European communities needs to be traced back to any particular prehistoric lineage, split out of 7 mitochondrial lineages (7 daughters of mitochondrial Eve) from the European continent, or out of the globally recognized 9 matriarchal. Mitochondrial lineages (Wade, May 2, 2000). From, business perspective, Sykes' business plan in this unexplored domain of applied genealogical investigation, based on mtDNA analysis, is found to be fantastic to its potential customers as an individual, belonging to European communities, would have a valid reason to be inquisitive to explore his or her genealogical identity in the mythological framework, that rendered determining his or her identity, like belonging to a specific haplogroup and tentative time of origin, after being descended out of any specific matriarchal lineage, out of 7 daughters of European Eves. Technically, DNA analyzing protocol, which was adopted by Syke's www.oxfordancestor.com has found to be a simple and straight-forward protocol, as it extracted the DNA sample from saliva from the mouth cavity of its clients by collecting through the buccal swab, and it charged only $180.00 for this test to its clients and the expenditure for this test has not found to be absurd, that made Sykes' genealogical investigating company popular to its potential clients. The genealogical identity of any extant representative of Sapiens, would help it to understand its matriarchal evolutionary root, suppose the analysis of the mtDNA of any individual, that helped the concerned individual to recognize belonging to any specific haplogroup "U," and helped him or her to accept as the successor of any prehistoric predecessor, descended from Ursula, who likely lived in the northern part of Greece around 45,000 years ago (Wade, May 2, 2000). Likewise, if the analysis of the mtDNA had pointed out any living representative of Sapiens, belonging to haplogroup X, supposedly recognized him or her further as the descendant of the prehistoric ancestral lineage of Xenia who lived in the Caucasus mountainous regions around 25,000 years ago. To make it a globally recognized, unique model for genealogical investigations, Professor Sykes had stepped out of his European model, to be engaged in investigating the mtDNA sample of any individual, across the globe, to establish its evolutionary identity to its prehistoric, matriarchal lineage by figuring out whether it belongs to any specific haplo-lineages out of 14 lineages in Africa or the 16 lineages in Eurasia and America (Wade, May 2, 2000).

To solve the mystery of Spatio-temporal origin and evolution of Sapiens, the phylogenomics analysis of Dr. Vanessa Hayes, Garavan Institute of Medical Research, and the faculty of the University of Sydney, Australia, needs to be referred here, which has been published in Nature on 18th October 2018, and in this publication a detailed account of genomic analysis of 1200 indigenous Africans (lived in the Southern part of Africa) has been narrated in details (Chan et al, 2019). Dr. Hayes's meticulous genomic analysis of these indigenous Africans, helped to understand and define further that their evolutionary identities have been traced back as descendants of the archaic DNA lineage, having identical mitochondrial genes of similar molecular signature, recognized as L0, inherited vertically for generations after generations from the prehistoric predecessors to its successors, as a result of maternal inheritance, and remained unaltered for hundreds of thousands of years (Chan et al, 2019). The most interesting part of the result of this investigation, carried out by Dr. Hayes and her research associates, had indicated towards the tentative time and place of origin and evolution of the ancestral lineage, L0 had supposedly taken place as these important findings have unfurled the history of origin and evolution of Sapiens on Earth. After critical reviewing the molecular signature of this genomic analysis, Dr. Hayes and her research colleagues shared the result of this unique experimental research, which has helped the paleoanthropologists and molecular phylogeneticists further to contemplate that L0, the prehistoric evolutionary predecessor of Sapiens, has thrived for hundreds of thousands of years till date since their maiden date of emergence on Earth (Specktor, October 28, 2019). Furthermore, this meticulous analysis on mtDNA of L0, prehistoric lineage of Sapiens, would inspire Dr. Hayes and her research colleagues to share their observation on the tentative time in that prehistoric time, whenever the main prehistoric archaic lineage L0 split out to form some sub-lineages, the existence of such lineages of descendants have anthropologically been substantiated by the presence of several small, extant, indigenous populations of Sapiens, confined in the African continent (Specktor, October 28, 2019). Based on her experimental observation on the genomic analysis on L0, Dr. Hayes and her research colleagues are very much convinced about the tentative time of origin of Sapiens on Earth, but their main concern has not been fully resolved as they could primarily determine the place of origin of Sapiens and it has geologically been recognized as a big continental mass, but their earnest endeavor to narrow it down on a geographical mass as the maiden birthplace of humanity has been remained unexplored yet. Hence, Dr. Hayes has claimed further "*we've known for a long time that humans originated in Africa and roughly 200,000 years ago...But what we hadn't known until this study was where, exactly this homeland was*" (Chan et al, 2019).

However, the molecular phylogeneticists, evolutionary biologists, and genomicists have started arguing that to unfurling the history of origin and evolution of Sapiens, a critical investigation with mtDNA is inevitable as mtDNA has found to be the key molecular tool in defining the maternal inheritance or matriarchal descendance of inheritance from archaic maternal predecessors to its extant successors. It appears that the presence of nuclear DNA is genetically and functionally more conspicuous than extra-nuclear or mtDNAs as mitochondrial DNAs have appeared to be a fraction of the human genome, as the human genome is the treasure-trove of nuclear DNA. Furthermore, a big reserve of nuclear DNA has been found in the cell nuclei and it has out-numbered the content of mitochondrial DNA reserves in the mitochondrial organelle in our cell. The reproduction biologists observed that inheritance of nuclear DNA in the offspring from both parents, which is accomplished through successful completion of sexual reproduction between a matured male and female individual of Sapiens, has usually gone through a cascade of genetic recombination and mutational changes in each generation. Whereas there is no such possibility of recombination and mutational changes whenever offsprings inherit mtDNA from its maternal lineage. It has also been observed by the human genomicists that inheritance of mtDNA from maternal lineage to its offspring have been remained unchanged for hundreds of thousands of years, as the rate of mutational changes in mtDNA is considerably very low as scientists have observed that the average rate of mutational changes in mtDNA is found to be 1 in 10,000 years and it happened in 300-600 generations cycle (Gibbons, 1998; Secktor, October 28, 2019). So, the human genomicists have reasonably claimed that to track down and narrate the evolutionary history of Sapiens which begun with prehistoric emergence of L0 lineage followed by evolutionary diversifications and split into sub-lineages have appeared to the contemporary derived lineages, the molecular footprints of mtDNA or mitogene have played an important role. Hence, exploring and ensuring the existence of L0 in the extant lineages of Sapiens has intrigued the geneticists to contemplate the fact that offspring have supposedly inherited the aforesaid mitochondrial gene or mitogene from their mothers and the typical mitogene has been descended through the matrilineal lineage (having no mutational changes) since it has emerged in the DNA sequence of mitochondrial Eve, who has also been recognized as the primordial woman or the universal mother of all mothers in Sapiens (Chan et al, 2019). After a long term and tedious investigations with molecular foot-print of mitogen, Dr. Hayes and her research colleagues had contemplated that a couple of indigenous groups of populations of Sapiens in South Africa, belonging to the Khoisan tribal communities, have been descended out of L0 matriarchal lineage. The next part of the investigation of Dr. Hayes is more interesting whenever they have critically reviewed the sublines (split away lines from L0 over time), those descended out of L0, found to be represented by a small number of indigenous communities, lived across the pan-African region. These unique findings had intrigued Dr. Hayes and her research colleagues to undertake an empirical comparison of evolutionary divergence by backtracking the comparison of evolutionary age between L0 and derive lineages of L0 by standardizing a molecular clock, and it helped to determine the tentative evolutionary age of divergence of L0 in African Eve (Chan et al, 2019).

To coordinate this part of the investigations, Dr. Hayes and her research colleagues had sequenced around 200 L0 mtDNAs collected from the individuals belonging to the small indigenous communities in South Africa. At the same time the researchers compared the database of 1,000 L0 sequences, to grasp an idea, how the archaic lineage of Sapiens would have gone through the trail of evolutionary diversifications to split away further and give rise to derived lines or subgroups of L0 with time, and found all over African continent; further ascertained by the possession of typical molecular foot-print of the mitogene in the mtDNA of the diverse indigenous communities in South Africa, originally found in the mtDNA of L0. Dr. Hayes and her research colleagues critically reviewed these experimental results and decided to contemplate as the preliminary, spatio-temporal milestone to narrate the beginning of the history of descendants of African Eve, supposedly evolved around 200,000 YBP, somewhere in the African continent and those matriarchal lineages had further split out in distinct sub-lineages, leading to the emergence of diverse indigenous groups across pan Africa, successively followed by the emergence of the modern communities in Sapiens, who then migrated and inhabited most of the terrestrial ecosystems on planet Earth (Chan et al, 2019). However, after a prolonged and meticulous effort of their genealogical investigations, Dr. Hayes and her co-researchers had claimed that they are in the right direction in the sacred soil of the prehistoric home-land of Mitochondrial Eve as Dr. Hayes has firmly claimed with confidence "*using that [evolutionary split of L0 subgroups from L0 haplogroup lineage], we could pinpoint what we believe is our human homeland*" (Chan et al, 2019). Though it seems to be an impossible task to most of the scientists, from interdisciplinary fields, ranging from molecular genetics to phylogenomics, to evolutionary biologists to paleoanthropologists, engaged in the long haul in exploring the birth-place of Sapiens as well as the home-land of mitochondrial Eve, but Dr. Hayes was found to be the ultimate winner of this marathon and she claimed that "Makgadikgadi-Okavango," a vast paleo wetland in South Africa of 120,000 square kilometers in area and Africa's largest palaeo-lake in South Africa, as well as in African continent, has appeared to be the maiden home-land of Sapiens; alternatively it would rather be recognized as the "Eden on the Earth" for Mitochondrial Eve. The unique result out of the genealogical investigations, made by Dr. Hayes and her co-researchers had narrated the prehistoric origin of Mitochondrial Eve supposedly took place at "Makgadikgadi-Okavango" in the southern part of Africa around 200,000 YBP, and seems to have stayed in the same location till 170,000 YBP, before stepping out 30,000 years to explore the uncharted territory in pan African regions, whenever L0 lineage branched-out of this main haplogroup to give rise to L0 subgroups (Specktor, October 28, 2019). Hence the research findings of Dr. Hayes has tried to firmly substantiate her observations that the ancient haplogroup L0 had stayed over in "Makgadikgadi-Okavango" paleo-wetland site for 30,000 YBP, unless the prehistoric community of Sapiens grew up to the size that might face the intraspecific struggle for food and space for survival and compelled to step out in course of first historic intra-zonal migration, anthropologically recognized as "Out of Africa" prehistoric event (Specktor, October 28, 2019). The maiden effort of intrazonal migration of archaic community of Sapiens, which has been recognized as "Out of Africa" has helped us to comprehend that there must be any convincible reason, like the disproportionate size of the community of the archaic community of Sapiens in respect to the availability of its dwelling space, it may be unavailability of food for all members of such archaic communities, it may be the disproportionate ratio of male to female individuals in the community, sudden alterations around "Makgadikgadi-Okavango" region, or an overlapping impact of two or more such reasons, rendered our prehistoric greatest grandmothers and grandfathers to leave behind their erstwhile homeland to explore uncharted territory out of Africa (Specktor, October 28, 2019). Based on the analysis of the geological sample, collected from the archaeological site of "Makgadikgadi-Okavango," Dr. Hayes and her co-researchers hypothesized a series of community actions like hunting, raising the families and other household engagements had rendered the prehistoric, indigenous communities of Sapiens to inhabit in the metaphoric Garden of Eden for 30,000 years, but our ancestors would have failed to cope with alteration of climate changes (supported by geological analysis of the sediments in "Makgadikgadi-Okavango"), supposedly taken place in "Makgadikgadi-Okavango" around 170,000 YBP, and that might have driven out our Great Grandparents out of the African continent (Chan et al, 2019). The hypothesis on the occurrence of sudden climatic changes in "Makgadikgadi-Okavango" has critically analyzed the sediment core samples from this area and running a paleoclimate-proxy and model data, and it led the paleo-anthropologists, meteorologists, and geologists to conclude that a substantial change of micro-environmental conditions of "Makgadikgadi-Okavango," especially the change of rainfall and enhanced level of humidity in this paleo-wetland area had likely opened up the green corridors towards north-east and south-west directions of this maiden birthplace of archaic Sapiens. Furthermore, Dr. Hayes had hypothesized that pre-historic, paleo-climatic changes which opened up the green corridors towards north-east and south-west directions of "Makgadikgadi-Okavango," through which the maiden intra-zonal migration, supposedly took the archaic group of Sapiens out of its metaphoric garden of "Eden" between 130,000 to 110,000 YBP (Chan et al, 2019). This maiden occurrence of intra-zonal migration of Sapiens out of African homeland has supposedly leading the L0 and its subgroups to disperse away from "Makgadikgadi-Okavango," to invade several uncharted territories to cope in different geo-climatic conditions and it helped the emergence of the descendants of L0 subgroups to form many small and new human settlements, formed by those indigenous communities of Sapiens leading existence of human communities of several L0 subgroups around this paleo-wetland area in South Africa, the metaphoric garden of "Eden" (Chan et al, 2019). As a model genealogical investigation, Dr. Hayes observation and interpretation on the maiden homeland issues of Mitochondrial Eve is found to be stunning but it has a noticeable downside, also as it has given a feasible explanation about recognition of the historical homeland of Mitochondrial Eve only, this unique genealogical investigation had failed to give any clue about the tentative time and place of origin and evolution of Y-chromosomal Adam. However, the mystery of origin and evolution of Y-chromosomal Adam or the counterpart of Mitochondrial Eve has tried to be elucidated by the scientists from the interdisciplinary field of researches like molecular phylogeneticists, evolutionary biologists, human genomicists, and paleoanthropologists, in a unique way as several scientists have claimed that the discovery of the historical homeland of African Eve, which has been defined by Dr. Hayes, might not be necessarily recognized as the same homeland for African Adam (Specktor, October 28, 2019). In support to the hypothesis on origin and evolution of Y chromosomal Adam, mentioned above, Dr. Chris Stringer of Natural history Museum, London, U.K. has expressed his concern with firm affirmation as he claimed "*Looking at the male-inherited Y chromosome, the most divergent lineages currently known in extant humans are found in West Africa. Not South Africa, suggesting our Y-chromosome ancestors may have originated from there*" (Specktor, October 28, 2019). After traversing through a labyrinth of possibilities, based on their findings on genealogical investigations, carried out by Dr. Stringer and Dr. Hayes, who had finally agreed to promulgate the ideas in favor of multiple homelands for African Eve and African Adam, and which has likely to be witnessed all over pan-Africa region, instead of designating any specific location in the African continent, and that might help us to contemplate the ideas of "metaphoric Garden of Eden," rather than promoting the concept of "metaphoric Garden of Eden" as the birthplace of Sapiens on the planet Earth (Garvan Institute of Medical Research, October 28, 2019; McKie, January 5, 2020).

References

Chan, E.K.F., Timmermann, A., Baldi, B.F., Moore, A.E., Lyons, R.J., Lee, S.S., Kalsbeek, A.M.F., Petersen, D.C., Rautenbach, H., Förtsch, H.E.A., Bornman, M.S.R., and Hayes, V.M. (2019). Human origins in a southern African palaeo-wetland and first migrations. *Nature*. 575(7781):185-189. doi: 10.1038/s41586-019-1714-1. Epub 2019 Oct 28. Erratum in: Nature. 2021 Apr;592(7852): E7. PMID: 31659339.

Elhaik, E., Tatarinova, T. V., Klyosov, A. A., and Graur, D. (2014). The 'extremely ancient' chromosome that isn't: a forensic bioinformatic investigation of Albert Perry's X-degenerate portion of the Y chromosome. Eur J Hum Genet. 2014 Sep;22(9):1111-6. doi: 10.1038/ejhg.2013.303. Epub 2014 Jan 22. PMID: 24448544; PMCID: PMC4135414.

Garvan Institute of Medical Research. (October 28, 2019). "The homeland of modern humans." ScienceDaily. ScienceDaily,

<www.sciencedaily. com/ releases/2019/10/191028175138.htm>.

Gibbons. A, (1998). Comparative genetics: Which of our genes make us human? *Science.* 281(5382): 1432-1434.

Gugliotta, G. (July, 2008). The Great Human Migration.

https://www. smithsonianmag.com/history/the-great-human-migration-13561/

Groucutt, H. S., Petraglia, M. D., Bailey, G., Scerri, E. M., Parton, A., Clark-Balzan, L., Jennings, R. P., Lewis, L., Blinkhorn, J., Drake, N. A., Breeze, P. S., Inglis, R. H., Devès, M. H., Meredith-Williams, M., Boivin, N., Thomas, M. G., & Scally, A. (2015). Rethinking the dispersal of Homo sapiens out of Africa. *Evolutionary anthropology*, *24*(4): 149–164.

https://doi.org/10.1002/evan.21455

Harari, Y.N. (2011). *Sapiens:A brief history of humankind.* Vintage Books, London. P.485.

McKie, R. (January 5, 2020). The Search for Eden: In pursuit of humanity's origin.

https://www.Theguardian.com/world/2020/jan/05/the-search-for- eden-in-pursuit-of-humanitys-origins.

Meyerhoff, J. (2013). *The Story of Us*. Lulu.com. p 66.

Specktor, B. (October 28, 2019). *Scientists think they have found Mitochondrial Eve's first homeland.*

https://www.livescience.com/mitochondrial-eve- first-human-homeland.html

Sykes, B. (2001). *Seven European Daughters of Eve.* W. W. Norton & Company Inc. p. 320.

Wade, N. (May 2, 2000). The Human Family Tree: 10 Adams and 18 Eves.

https://www.Nytimes.com/2000/05/02/science/the-human-family-tree-10-adams-and-18-eves.html.

Zimmer, C. (August 23, 2004). Adam and his Eves. https://www. nationalgeographic.com/science/article/adam-and-his-eves

Chapter 12

Delving for Metaphoric Garden of Eden: Humanity's First Homeland on Planet Earth – II

In fact, their [evolutionary biologists] hypothesis creates a sort of space-time paradox whereby the most ancient individual belonging to Homo sapiens species has not yet been born. If we take the numerical results from previous studies seriously we can conclude that the past may be altered by the mother of Adam deciding not to conceive him in the future, thus, bringing a retroactive end to our species.

---- Eran Elhaik

"Where will your DNA take you?... Learn about your personal history and follow the path of your ancestors with Family Tree DNA's industry-leading tests..." It appears to be one of the most persuasive taglines for any individual of Sapiens, used by "Family Tree DNA," one of the leading US-based industries in the domain of molecular genetics and genomics, specifically engaged in genealogical analysis. Most likely, the eye-catching commercial of "Family tree DNA" has intrigued one of the close female relative of Albert Perry, most recently deceased, African-American, who used to live in South Carolina, to be inquisitive about Perry's evolutionary descendants as well as identity concerning any specific haplogroup or archaic lineages of Sapiens around hundreds of thousands of years ago, and that made the female relative of Perry deposit her DNA sample to "Family Tree DNA" for genealogical analysis. The molecular use such DNA sample to analyze our family tree to reveal how we relate to our ancestors and US-based "Family Tree DNA," the genealogical analysis laboratory has a proven track record of successfully analyzing the DNA sample of their hundreds of thousands of clients. Though analysis of each DNA sample has appeared with unique results that helped every client to recognize their prehistoric ancestral lineages (or specific haplogroups), they are phylogenetically related to but geneticists have witnessed that there are two key findings, observed in those analyses (Barras, March 6, 2013):

1. All male individuals of Sapiens have inherited Y-chromosome from the prehistoric, common, male predecessor, should be scientifically recognized as genetic "Adam" or "Y-chromosomal Adam" and,
2. The calibration of the molecular clock of Sapiens has revealed that genetic "Adam" or the "Y-chromosomal Adam" had supposedly evolved and lived between 60,000-140,000 YBP on the Earth.

The DNA technologists of "Family Tree DNA" tried to analyze the DNA sample of Albert Perry to place Perry on the standard format of Y-chromosomal family tree but their preliminary effort has remained unsuccessful as Perry's Y chromosome has appeared to be a quite unusual appearance, in terms of molecular genetics and it has not resembled any of the DNA samples of hundreds of thousands individual of sapiens earlier analyzed. The preliminary observation on the molecular configuration of Perry's Y-chromosome has simply amazed the molecular geneticists and they claimed that it has found to be clearer distinctive among all DNA sample, been analyzed so far, so it would inevitably come up with unique observations, to understand the origin and evolution of Sapiens on Earth. Furthermore, the preliminary level of DNA analysis report has rendered the molecular phylogeneticists to contemplate that the evolutionary age of the "last common male ancestor" of our parental lineage of archaic Sapiens appeared to be more than two times, as it has appeared earlier (Barras, March 6, 2013). To give feasible explanations, the genomicists and anthropologists have tried to promote the "Interbreeding hypothesis of archaic Sapiens" as they have promulgated that around hundreds of thousands of years ago, the archaic members of Sapiens have engaged with interbreeding course of actions with the individuals of anatomically modern members of Sapiens somewhere in Central Africa and the scientists have further referred such incidental occurrence of interbreeding between archaic Sapiens and Denisovans in Southeast Asia and the incidental occurrence of interbreeding between archaic Sapiens and Neanderthals in the Middle East Asia (Barras, March 6, 2013).

However, the scientists and DNA technologists of "Family tree DNA" have failed to match or compare Allen Perry's Y-chromosome to any chromosomal profile of any individual of the Sapiens analyzed so far, to fit in the universal template of the Y-Chromosomal family tree of Sapiens. It intrigued Michael Hammer one of the renowned scientists from the University of Arizona, Tuscon, USA, and his research colleagues to take the challenge to undertake a study on analyzing the unusual Y-chromosome of Allen Perry. Based on their critical reviews, Michael Hammer and his research team claimed that the distinct but unusual Y-chromosome of Allen Perry has likely to be separated from other lineages of Sapiens around 338,000 YBP. Based on the age-dating protocol, the paleo-anthropological investigations have ascertained that the anatomically modern descendant of Sapiens supposedly evolved around 195,000 YBP; so it is wondered the molecular geneticists to contemplate that Allen Perry's archaic lineage had split from the rest of the evolutionary lineages of Sapiens much earlier during the emergence of *Homo sapiens* on Earth. In a different experimental observation on the tentative time-frame of the emergence of anatomically modern species of Sapiens, Dr. Eran Elhaik has observed and shared the identical view along with other scientists as he said "*We can say with some certainty that modern humans emerged in Africa a little over 200,000 years ago*" (Elhaik et al, 2014). Furthermore, they have made a radical statement that Allen Perry did not descended from "genetic Adam" (Barras, March 6, 2013). Going further, Dr. Michael Hammer and his research colleagues have reviewed the Y-chromosome database, made on 6000 African individuals of Sapiens and surprisingly noticed that the Y-chromosomal profile of Allen Perry was found to have closely resembled another 11 men, who used to live in one of the villages of Cameroon (Barras, March 6, 2013). Such unique findings helped the molecular geneticists and evolutionary biologists to have an idea of the prehistoric location of origin of Allen Perry in the African continent. In this regard, the review researches of some contemporary scientists, those who have gone through some research findings on the evolutionary occurrence of Y-chromosome of Allen Perry needs to be shared here. Dr. Chris Tyler-Smith at Wellmore Trust Sanger Institute in Hinxton, U.K., has commented that he does not doubt to claim that the evolutionary age of Perry's Y-chromosome seems to be much older, than the conventional perception we have on the prehistoric time of emergence of Y-chromosome of Sapiens. Dr. Jon Wilkins of Roin Institute of Montclair, New Jersey, USA, made a unique observation on the experimental observations on Perry's unusual Y-chromosomal profile and he claimed "*It is a cool discovery...we geneticists have been looking at Y chromosomes for about as long as we've been looking at anything. Changing where the root of the Y-chromosome tree is at this point is extremely surprising*" (Barras, March 6, 2013).

However, one of the feasible hypotheses, nurtured by the evolutionary biologists and molecular geneticists has intended to present the inheritance of the genetically unique Y-chromosome by Allen Perry was most likely accomplished from the archaic, ancient, African Sapiens ancestral lineage, that likely was pushed into extinction since then. It has further been extrapolated by the evolutionary biologists that the archaic genome of such archaic African Sapiens was likely to be engrossed as a result of interbreeding between them and the populations of anatomically modern species of Sapiens (Barras, March 6 2013). Later, the evolutionary biologists have referred to some interbreeding scenarios between the archaic lineage of Sapiens and the anatomically modern human species, which helped the scientists to contemplate the genomic introgression of the archaic lineage of Sapiens in the representatives of the modern, anatomically advanced member of Sapiens. In the year 2000, some molecular scientists have critically investigated the human fossils, previously discovered in the fossil site of Iwo Eluru in Nigeria, and the scientists have witnessed a magnificent incident, as they have noticed that the fossil bodies of the prehistoric Sapiens showed composite features of the Archaic Sapiens and the anatomically modern Sapiens, and that has indicated the strong possibilities of the interbreeding between the Archaic and modern lineage of Sapiens. Such experimental interpretations have inspired Micheale Hammer of the University of Arizona has stated with affirmation that "*The Cameroon village with an unusual genetic signature is right on the border with Nigeria, and Iwo Eleru is not too far away*" (Barras, March 6. 2013). Dr. Chris Stringer, at Natural History Museum, London, UK, who has been involved in the genetic data analysis of the fossil specimens, from Iwo Eluru, Nigeria, has expressed his concern that more genetic data are required from Sub-Saharan African populations of archaic Sapiens. Furthermore, Dr. Stringer has claimed observed: "*The oldest known fossil humans both in West Africa at Iwo Eluru [in Nigeria] and Central Africa at Ishango [in the Democratic Republic of Congo] show unexpectedly archaic features, so it certainly looks like we have a more complex scenario for the evolution of modern humans in Africa.*" So, such critical observation of Dr. Stringer has not only ushered discarding the earlier claim of "Makgadikgadi Okavango," at Southern part of Africa and which has been contemplated by the scientists as hypothetical birthplace as well as metaphoric Garden of "Eden" for Mitochondrial Eve, might not necessarily be recognized to be the birthplace of Genetic Adam or "Y-MRCA Adam." Furthermore, Dr. Stringer's observation on the archaic members of Sapiens at West, Central and Southern part of Africa, has promoted another hypothesis of origin and evolution of Sapiens, that claims that prehistoric emergence of Sapiens had supposedly happened more than once in the wide Spatio-temporal backdrop, somewhere in the African continent.

In this regard, the experimental observation of Dr. Eran Elhaik from the University of Sheffield, U.K., and his research associates on the biological models of Y-Chromosomal Adam need to be mentioned, as DR. Elhaik and his associates claimed (based on their analysis on Y-chromosomal DNA) have claimed that from an evolutionary perspective, Y-chromosomal Adam was found to be 8,300 years older than scientists originally believed (Elhaik et al, 2014). Dr. Elhaik and his associates tried to figure out whether overlapping of the evolutionary emergence timeframe of Y-chromosomal Adam and Mitochondrial Eve is scientifically convincible or not. The contemporary research findings on Y-chromosomal DNA of Sapiens, which was recently published in "American Journal of Human Genetics" in February 2013, Dr. Elhaik and his research colleagues claimed that the origin of Y-chromosome in Sapiens took place as a result of interbreeding and emerged in a different species of archaic Sapiens, who had supposedly almost twice elder than "Genetic Adam" (Elhaik et al, 2014). It is a matter of concern to molecular geneticists and evolutionary geneticists to nurture some hypothetical situations, like theoretical probabilities of interbreeding between the archaic lineage of Sapiens, and the recently evolved anatomically modern human species, followed by the genomic introgression of the archaic lineage of Sapiens into the derived genome of newly evolved anatomically modern representative of Sapiens. The evolutionary biologists and genealogists further expressed their skepticism to such theoretical probabilities of interbreeding between the modern representative of hominins and their archaic predecessors, supposedly evolved around 500,000 YBP. Furthermore, the theoretical probabilities of the emergence of archaic lineages of hominins in a different place, in different period of the prehistoric timescale of biological evolution, has promoted the ideas of multiple times of origins and evolution of hominins on the Earth and further extrapolation of such hypothesis helped to postulate the emergence of more than one mitochondrial Eves and more than one genetic Adams, our prehistoric predecessors, who have gone through the reticulate evolutionary history of origin and evolution of hominins on Earth for a long time, to have emerged as modern human species. But the advanced level of genealogical investigations in quest of defining the origin and evolution of modern representative of Sapiens, the current researches would not discard the possibilities of interbreeding between the archaic and modern hominins and genetic introgression of archaic genomic molecules in the genome of the modern hominins. It has remained a challenge to the genomicists, molecular geneticists, and genealogists that up to what extent such probabilities of interbreeding between such distinct group of hominins are possible as it has been echoed in the statement of Dr. Elhaik, who commented: "... *the question to what extent did our human forebearers interbreed with their closest relatives is one of the hottest questions in anthropology that remains open*" (Elhaik et al, 2014).

The critical analysis of palaeontological pieces of evidence of ancient human fossils, along with palaeo-anthropological evidence of their tools, articles of clothing, utensils, pieces of jewelry, etc., used by the archaic communities of Sapiens in prehistoric time; supplemented by the examination of molecular footprints of them, followed by the intensive analysis of genealogical pieces of evidence helped the researchers to put together to define the history of origin and evolution of Sapiens. Since its emergence on Earth, some inter-disciplinary researches, mentioned, above, done by the scientists from time to time had rendered the scientists to promulgate that the extant genealogical lineages of anatomically modern descendants of Sapiens, likely to be emerged out of the evolutionary cradle of archaic root-stock of Sapiens, the "hunter-gatherers," supposedly roaming around in African continent before stepping out of its maiden homeland to explore the rest of the world.

Though there are arguments and counter-arguments of the evolutionary scientists, engaged in the exploration of the exact birthplace of Hominins and the quest for the metaphoric Garden of Eden is still one as there is no unanimous resolution on it, to admit by all scientists that all or most of them would recognize any specific location as the sacred birthplace of hominins, as the genealogical analyses on the emergence of Mitochondrial Eve and Y-chromosomal Adam have projected more than one geographical location and different period of geological time scale to conclude any specific space and time as the evolutionary birthplace of hominins in a prehistoric time. But the collective observation of the interdisciplinary scientists have helped them to come up with their composite findings to contemplate as African continent as the sacred birthplace of our ancestors' tens of thousands of years ago, and it started haunting our minds to think about our prehistoric origin and identity where and when it had happened? When a group of scientific analysts, having expertise in molecular genetics have firmly claimed that the origin and evolution of Hominins likely took place, either in Ethiopia or Kenya of Eastern Africa, another group of evolutionary biologists have prioritized their genealogical analysis and tried to substantiate their claim in favor of South Africa as the mystic Garden of Eden, and amid such ongoing bickering's of claims and counter-claims, the third group of human geneticists has found to be quite optimistic to identify the exact location or more than one locations of Eden, as their observation and elucidation of experimental data and its interpretation for the long-hauled, sublime and mystic abode of Sapiens would likely to end the term "evolutionary standoff," that we, the anatomically modern representative of hominins have gone through for a couple of centuries.

If and when we would accept the principles of biological evolution is the ultimate impetus for origin and evolution of Sapiens on Earth, the evolutionary birthplace of Sapiens has been perceived like strange and sublime "Stone-age paradise," where the contemporary version of Sapiens, the omnipotent lord of Anthropocene on Earth has come across the labyrinth of the evolutionary trail of history of the origin of *Homo sapiens*, from their prehistoric cradle of archaic predecessors to inhabit this stone age paradise, also known as the mystic garden of "Eden" for playing around the metaphoric apple-orchards. The contemporary representative of hominins, the metaphoric lord in Anthropocene, has undoubtedly traversed a long trail of biological evolution since its emergence from the archaic cradle of Sapiens, around hundreds of thousands of years ago and the anatomically modern representative of Sapiens has undoubtedly inherited the characteristic features of a hominin from their archaic ancestors, but there is no doubt that it has gone through the rigorous screening of natural selection in the backdrop of the ever-changing environment of this biosphere, that helped this version of Sapiens to be emerged with a glancing potential of intellects, along with the highest extent of behavioral, cultural and cognitive skills as the living biological legend on the planet Earth. When we try to figure out our identity as a modern representative of hominins, there are two paths normally we choose to establish our ethnicity, either by visual matching by judging the degrees of visual similarities or dissimilarities of us about our archaic ancestors or trying to figure out the apparent location of our homelands on the geographical map. There is an artificial system of identification (racial profiling) found to be existing and popularly practiced in diverse geopolitical regions to identify the largely-based communities of modern hominins (popularly recognized as nations) and it has been practiced mainly based on the key features: structural orientation, like the existence of variations of melanocytes, under the subcutaneous layer of our skin, the color of our iris in our eyeballs, and color of our hair, etc. But the matter of fact is that such superficial categorization of 1.6 billion modern Sapiens (in the conventional template of racial profiling) is found to be pseudo-scientific to reveal our identity. In the backdrop of evolutionary genetics, we could not denounce the hard reality of our origin and evolution that helped us to traverse a long trail of biological evolution, to emerge as an anatomically modern representative of Sapiens, once that journey had begun around hundreds of thousands of years ago, somewhere across the African continent. Our genealogical identity has been buried in the past, which needs to be unearthed to acknowledge the truth that we are all descendants of our African ancestors, the great and great-grandsons and grand-daughters of "Mitochondrial Eve" and "Y-chromosomal Adam," our sublime, metaphoric grandparents of humanity on planet Earth.

Most of us love to dream a panoramic picturesque of arcadia when it appears to be a long-cherished haul for the metaphoric garden of "Eden" in our mind, as the evolutionary birthplace of humanity appears to be serendipitous and sublime location, touching down of that sacred place which seems to be our ultimate pilgrimage, to offer prayer to our archaic predecessors to be recognized as well as identified as the contemporary descendants of hose valiant lineage of Sapiens, dared to fight against all odds in their ambiance in the race of survival that we, the modern representative of hominins inhabit in this congenial abode of planet Earth in peace. It is a matter of fact that our evolutionary scientists (comprising of paleo-anthropologists, archaeologists, evolutionary biologists, etc.), engaged in traditional research to unfurl the evolutionary history of hominins, have remained unsuccessful for a long period, to explore as well as identify any specific geographical location as the maiden birthplace of Sapiens on Earth, mainly due to combined factors of unsuccessful attempts and lack of tenacity. The combined bottleneck of lack of tenacity and unsuccessful attempts of exploration of the maiden birthplace of humanity on Earth has enhanced the level of skepticism in next-generation scientists (comprised of molecular geneticists, genealogists, genomicists, molecular phylogeneticists, and evolutionary geneticists, etc.) engaged in the evolution of hominins on Earths and their expression of skepticism have resonated in the words of Dr. David Reich, one of the famous geneticist from Harvard University, USA, as he has firmly stated: "*When it comes to human ancestry, there was no Garden of Eden*" (McKie, January 5, 2020). Whenever the meticulous efforts of the paleoanthropologists, paleontologists, archaeologists, molecular geneticists, human genomicists, evolutionary biologists, and other interdisciplinary scientists have reached a dead end, that failed to shed any light on the evolutionary history of Sapiens; the next-generation researchers have adopted a less-known path to trace-out the foot-prints of the archaic members of Hominins on Earth, and their narratives on the analysis of research findings have been able to deliver feasible explanations on our evolutionary perception of the prehistoric homeland in Africa. In an alternative narration, the evolutionary biologists claimed that the emergence of anatomically modern representatives of Sapiens out of the archaic cradle of Hominins, in a place in a particular time, seems to be most unlikely, rather they have promoted that evolution of modern hominins has supposedly taken place for more than once, that supposedly ended in for a period of half million years in the African continent. The narratives of Professor Chris Stringer, one of the renowned palaeoanthropologist and the visiting professor of University College London, Archaeology, U.K., has firmly supported the explanation on the evolutionary emergence of hominins in Africa, mentioned above, and he has further claimed that "…immediate predecessor of modern humans probably arose in Africa about 500,000 years ago and evolved into separate populations" (McKie, January 5, 2020). Based on his research, made on this topic of origin and evolution of anatomically modern Sapiens for last two decades, Professor Stringer has shared his findings, as it has been hypothesized that in the adverse time of paleoclimatic regime, whenever the environmental ambiance had rendered the Saharan region in Africa as the dry and desiccated zone of the desert, the isolated, small populations of ancestral human species had likely to struggle for its existence to be in the race of survival and it had likely pushed a big number of the human populations to the brink of extinction. Eventually, a small number of archaic human populations have ensured their survival in some pocket refugia like an oasis in the deserts in the middle of such adverse environmental regimes, and their status has metaphorically resembled the "living fossils." Hence, the reversal of such adverse environmental conditions in that paleo-climatic regimes had rendered an opportunity to those archaic community of human "living fossils," the ultimate survivors of the fragmented populations of archaic Sapiens, to come back in the rhythm of full-blown life and convincingly they had managed to increase population size, once substantially shrunk, and the evolutionary biologists had contemplated further that the rejuvenated archaic, indigenous populations of Sapiens had supposedly engaged in the random exchange of genetic materials in intra-specific level, once they engaged in intra-zonal migrations, out of its maiden birthplace, for exploration of the inhabitable, uncharted territory on planet Earth (McKie, January 5, 2020).

However, Prof. Stringer had observed that random occurrence of several physicals, physiological, genetic exchanges, genomic integrations have been integrated with intra-zonal migrations in the backdrop of paleoclimatic changes for the last 400,000 years of the overlapping evolutionary history of hominoids and hominids, have not only helped the global communities of Sapiens, to go through the exit of natural selection to have emerged in the form of anatomically modern representative of Sapiens in this regime of Anthropocene; it has helped the hominins to be globally well-diversified in terms of various degrees of genotypic and phenotypic profile and help us occupying all inhabitable niches of the global terrestrial ecosystems.

Further moving forward, Dr. Eleanor Scerri, one of the leading scientist on human evolution, at Max Planck Institute, Jena, Germany has presented the narrative of the evolutionary journey of modern human species on Earth in her research model, entitled "mix-n-match building blocks the evolution of hominin" in which she has portrayed that the anatomically modern human species are the vertical descendants, emerged out of the rootstock of a genetically interlinked group of Sapiens, often split out to form a new haplogroup and or merge to form a derived lineage of Sapiens. So, in the narrative of Dr. Scerri, the emergence of the anatomically modern human species happened in distinct populations in different regions of the world, and the impact of an admixture of physical, physiological, and genetic principles and an amalgamation of cross-cultural habits helped to promote the formation of a mix-n-match profile of Sapiens, possessing with ancestral and modern traits, inherited as well as accumulated vertically from their immediate and distant evolutionary cousins for hundreds of thousands years (McKie, January 5, 2020).

The scientists have observed that the animal species, having a large population size and widespread distributional potential over a large geographical region, was found to be in the mode of expansion in a vast geographical area over time but failed to cope with the vortex of environmental changes over time, the erstwhile large-sized population of that animal would go through the cascade of population fragmentation and emerged into a number of sub-species. Over time these newly evolved sub-species are found to be acclimatized in the newly altered micro-environmental atmosphere and all these sub/infra-species/varieties, have gone through the trail of biological micro-evolution, that rendered the emergence of all these sub-species to get a foot-hold in new species. In the world of science, specifically in bio-science, there are several theories, found to be stunning, but it was not found to be compatible when scientists tried to examine whether it is compatible on the biological system in reality or not. The story of the formation of sub/infra-species in the biological world and its emergence as new species, in course of evolution, the theoretical probability of such occurrence on the biological system was found to be mind-boggling; in reality, it was not found to be so intriguing, when it has been implicated to define the origin and evolution of modern version of Sapiens out of its archaic predecessor. Fortunately, the evolutionary scientists observed that the rate of evolution of hominins has not complied with the theoretical perception of evolution for the animals, occupying a very large geographical space and found to be in the exponential model of expansion, so the frequency of fragmentation of the population of Sapiens is most likely. If the evolutionary progress of hominins has followed the classical theory of evolution, which has been normally been followed by the other animals (having a large size population and occupied a large geographical location) there are possibilities of emergence of many infra-species or varieties and the random occurrence of cross-fertilization and tentative possibilities of inbreeding might create a chaotic situation to define the history of origin and evolution of anatomically modern representative of sapiens. The evolutionary biologists have observed that the theoretical possibilities of fragmentation of modern hominin populations and the possibilities of emergence of some sub-species/varieties followed by their journey through micro-evolutionary changes, to be evolved as newly formed species, are less likely in reality for the modern representative of Sapiens and the scientists have contemplated that the "long-range social networking" spree of anatomically modern human beings has likely slowed down the pace of human evolution but gradually favored its emergence "en masse across Africa" (McKie, January 5, 2020).

Hence the evolutionary biologists who have tried to define and re-define the slow but gradual mode of origin and evolution of the anatomically modern representatives on Earth, have considered that "long-range social networking" should be considered to be one of the important behavioral and cognitive traits of the current version of Sapiens, that has played the key role in the evolution of hominins. In favor of the views, mentioned above, Professor Mark Thomas of University College London, U.K., one of the renowned molecular geneticists in Europe, strongly argued that its accumulation of a set of inter-related factors like culture, knowledge, belief, and its recognition of social values have played key roles in shaping our evolution to a certainty by ensuring our sustainability, when the indigenous or tribal populations across the African continent are found to be engaged in evolutionary diversification by action as the evolutionary baseline of hominins in the dawn of a civilization of hominins (McKie, January 5, 2020). Professor Thomas was confident in his observation as he has claimed that to achieve sustainability in the biological evolution of a social animal like the modern representative of Sapiens, cultural inheritance of behavioral and cognitive traits from the ancestors to its descendants was found to be one of the key factors, that prompted defining the part of the sustainable evolutionary history of Hominins and in support of his view, Professor Thomas has stated with the affirmation "*Without culture, we would be dead....We know things today that we were worked out by ancestors tens of thousands of years ago and have been passed along over the generations. Culture is our life-support system*" (McKie, January 5, 2020).

To understand the origin and evolutionary history of Sapiens, the experimental observation and interpretation of the contemporary investigations on the DNA sequences from Y-chromosome have helped Dr. Fernando L. Mendez, of Stanford University, USA, one of the leading population geneticist from North America and his colleagues have primarily noticed that existence of Y chromosome has exposed maleness in humans. Dr. Mendez and his research colleagues have published their work in the American Journal of Human Genetics, and their earnest endeavor to calibrate the molecular clock to review the evolutionary age of Y-chromosomal Adam or African Adam and their analysis prompted them to hypothesize that the African Adam is the most recent common ancestor (MRCA) of all males in the Sapiens, evolved around hundreds of thousands of years ago (Mendez et al, 2013). The critical analysis of the structural and biochemical configuration of hominin's DNA molecules, that has been published by Mendez and his colleagues have elucidated that there is a lot of variations, witnessed in the DNA molecules of the Y-chromosome and the extent of has found to be in ever-increasing mode over time (Mendez et al, 2013). The critical analysis of the molecular data, encrypted in the Y-chromosomal DNA of African Adam has made the scientists hypothesize that emergence of Y-chromosomal Adam or African Adam had supposedly taken place in the Western part of Africa (Mendez et al, 2013). Let's review the unique research work of Dr. Allan Wilson, of the University of California, Berkley, USA, one of the most eminent molecular biologists along with a group of young research scientists have analyzed the molecular footprints of mitochondrial DNA, and compared its variations, collected from the diverse group of global populations (McKie, January 5, 2020). The experimental research of Dr. Wilson has primarily revealed that the anatomically modern representative of sapiens emerged out of the archaic matriarch lineage from African Continent, popularly recognized as Mitochondrial Eve or African Eve. Hence, the research findings of Dr. Wilson have indicated further that the rootstock of the emergence of a large group of the modern version of Hominins, was likely initiated somewhere in Africa (supposedly Botswana) (McKie, January 5, 2020). The evolutionary biologists, those engaged in the exploration of the history of origin and evolution of the anatomically modern human species, have tried to demonstrate a happy-ending loves-story of African Adam and African Eve in the mystic apple-orchard, likely to be inhabited in the metaphoric garden of "Eden" and engaged in planning for of a bigger family. In reality, the evolutionary history of the modern version of hominin has not necessarily complied with the happy-ending narrative of Adam and Eve love-story at the Eden Garden, contrived by the evolutionary scientists, as the careful genealogical investigations on Mitochondrial Eve and Y-chromosomal

Adam have revealed that origin of African Eve and African Adam has not been found in different paleo-geographical locations, both of these distinct archaic lineages supposedly emerged in a different paleo-temporal regime of prehistoric time. Furthermore, the most disappointing part of the contemplated love story of African Eve and African Adam, portrayed by the evolutionary scientists, has lied in the emergence of African Eve and African Adam more than once in multiple locations in the African continent in hundreds of thousands of times and that part of the twist in this love-story has rendered the scientists to check any possibilities of prehistoric, spatial emergence of multiple African Eves and multiple African Adams across the African continent before they have engaged in crossbreeding, inbreeding, and intrazonal migration, to explore in the uncharted territories out of Africa.

In 1997-98, a group of paleo-anthropologists undertook a fossil expedition in the Blombos cave, South Africa. The scientists analyzed the artifacts and human fossil specimens, dispersed inside the cave. The critical analysis and age-dating process of the paleontological evidence from Blombos cave rendered the paleoanthropologists to contemplate that the geologically archaic lineage of Sapiens likely emerged in this geographical location, between 100,000 to 70,000 YBP (Grine et al, 2000). Furthermore, the experimental observation of the paleoanthropologists and the molecular geneticists have helped them to claim that the archaic hominin inhabitants in and around the location of Blombos cave had been descended out of Neanderthal rootstock in the Middle Stone Age (Grine et al, 2000). Donald Johanson, from Cleveland Museum of Natural History, USA, one of the famous paleoanthropologist of North America, discovered the fossil evidence of the prehistoric feminine entity, Lucy (AL288-1) from Hadar, Ethiopia in 1971, critically analyzed it and shared his important observations, in which he claimed that around forty percent of the archaic female representatives of Sapiens had likely descended out of *Australopithecus afarensis*, the anthropologically, archaic predecessor of the Sapiens, who supposedly evolved around 3.2 MYA (Johanson and Wong, 2009). Though age-dating process (popularly known as carbon-dating protocol) is one of the touch-stone technology, randomly employed by the paleontologists, to determine the evolutionary age of the fossil specimen, and it helped to compare the age of one fossil specimen, discovered from a specific geographical region, in comparison to its identical form in another specimen, discovered from another geographical location, yet it is found to be difficult for the paleontologists to ensure that evolutionary emergence of any specific archaic lineage has happened from any specific location, in a specific period. In respect to the paleontological observation and its interpretation, some genealogical analyses have tried to project that evolutionary emergence of African Eve and African Adam supposedly emerged from the African continent and both of them have likely lived in the same paleo-temporal regime and it has resonated the metaphoric fairy-tale of Adam and Even in the sublime apple-orchard of heaven, the metaphoric garden of "Eden" (Cann, 2013). The unusual twist in the genealogical interpretation of the fairy-tale depiction of Adam and Eve in the Garden of "Eden," has appeared in the narrative of the recent genealogical investigations on the archaic T-haplogroup chromosome, investigated by Dr. Elhaik of University of Sheffield, UK, and his research colleagues once they have published their observation on African Eve and African Adam in the European Journal of Human Genetics in 2014, as they have projected African Adam around 8300 older than the evolutionary age, projected by the conventional genealogical research models (Elhaik et al, 2014). Furthermore, the interpretation of the genealogical investigations, made by Dr. Elhaik and his colleagues, have claimed further that African Adam likely descended out of a different species of Hominins (supposedly accomplished in course of natural interbreeding) and that has projected the evolutionary age of the African Adam as almost double of the age, determined in the earlier genealogical investigations (Elhaik et al, 2014). The most intriguing narrative of the genealogical investigations, made by Dr. Elhaik, had presented a different perspective of the emergence of metaphoric Adam and Eve in the African garden of "Eden" as the researchers have claimed with a strong voice of affirmed that we, the anatomical descendants of Sapiens are the progenies of more than one African Eves and African Adams and supposedly evolved more than once in more than one locations across the African continent (Elhaik et al, 2014). Hence, the narrative of Dr. Elhaik had denounced the utopian evolutionary origin of hominins out of the archaic predecessors African Eve and African Adam in a particular period in a particular place, recognized as the metaphoric garden of "Eden" (Elhaik et al, 2014).

Hence, the contemporary genealogical investigations and its interpretations of Dr. Elhaik and his research colleagues have floated a new perception on the origin and evolution of the modern version of Sapiens as the slow but gradual evolutionary accumulative process on this biosphere. As it should not be considered as one-time emergence of hominins in the backdrop of defined Spatio-temporal reference, rather it would be considered as a gradual and composite evolutionary cascade, where the emergence of hominins should rather be considered as the outcome of evolutionary diversifications as an evolution of anatomically modern human species in this regime of Anthropocene has not appeared with it. The modern human species was not emerged with all its genetical, biological, and cultural traits in one shot as the paleo-anthropologists and evolutionary geneticists have studied the genotypic and phenotypic profile of some transitional archaic lineages of hominins and their descendants and defined that the transitional lineages of Sapiens have supposedly gone through a complex cascade of cross and interbreeding cycles in intra and inter-populational level, evolved and existed in the pre-historic time, that rendered the emergence of descended progenies of Sapiens with unique physical, physiological, genetical, and bio-behavioral traits, accumulation, and engrossment of unique genes in its genome in course of vertical and horizontal gene transfers. Gradual acquisition of unique genes in the gene pool of the modern human representative of Sapiens and intermittent process of genetic introgression had likely to expediting the extinction of the archaic lineages and the emergence of the derived lineages of Sapiens with unique genetic and bio-cognitive traits. The exponential expansion of the social network of the modern human species had been expedited by the prehistoric endeavor of the isolated, archaic communities,that brought more than one ethnic human communities close to each other, to be evolved as genetically versatile and anatomically modern human communities, likely to occupy the entire inheritable terrestrial ecosystems on the Earth. So, this unique evolutionary journey of modern hominins have rendered portrayal of a unique evolutionary family tree, the archaic main-trunk of that branched out profusely to represent some diverse human races, contending unique genetic principles; hence an effort of coalescing of these group of contemporary, derived lineages (a diverse group of human races) would likely help out tracking the root of humanity on Earth.

Anthropologists have considered that cultural inheritance is the key factor that the modern version of Sapiens have descended from their archaic predecessors, as the absence of cultural inheritance in the history of the evolutionary journey of anatomically modern human species would have rendered the story of origin and evolution of human, specifically the absence of scientific narrative of the behavioral evolution of this ultimate mammal in Anthropocene in this biosphere seems to transform it to a simple and conventional story of origin and biological evolution of any primates on Earth. So, it intrigued the evolutionary biologists like Chris stringer to recognize the behavioral evolution as the key impetus to narrate the origin and evolution of the modern hominins as he has firmly claimed "*There is no evidence at all that one single part of Africa produced all this modern behavior*," it makes sense (McKie, January 5, 2020). John Hawks, one of the famous anthropologist from the University of Wisconsin-Madison, USA, has expressed his concern on the contemporary, conventional "evolutionary tree models" as he has considered that these conventional models have been demonstrated as the theoretical metaphor in narrating the history of a modern version of hominins, but these models are far from reality in elucidating the history of origin and evolution humanity, rather he has promulgated an alternative analogy of "braided analogy," where the evolution of modern human species has resembled the multi-braided metal rope (normally used in the cranes), analogically could be referred as multi-channel rivers and rivulets, go through a labyrinth of merging and split for a time of hundreds of thousands years before emerged out as a huge water channel, evolved like a keystone water body in a vast geographical area (Hawks and Cochran, 2006). John Hawk's "Multi-braided stream" is a scientifically convincible metaphor that seems to narrate the origin and evolution of anatomically modern human species on the Earth.

The extensive level of investigations made on the molecular markers on an anatomically modern representative of Sapiens for last two decades rendered Dr. John Hawks to promote in favor of "Mix-n-Match" one of his favorite hypothetical promulgation, that claimed the tentative origins and evolution of the anatomically modern representative of Sapiens in more than one geographical locations, on more than one time. However, it needs to be addressed that the "Mix-n-match" hypothesis has originally been propounded by Dr. Chris Stringer, which has intrigued Dr. Hawks to accept it and advocate as one of the unique evolutionary narratives of origin of humanity and Dr. Hawks has come up with radical comments further "*...evidence began to emerge in 2010, showing that most living humans carry at least a small fraction of Neanderthal genes. Later that year, the Denisovan genome confirmed that some living human populations have* *genetic ancestry in multiple archaic human populations not yet discovered from archaeological or fossil evidence. But this year, the reality of population mixing blossomed, as ancient genomes themselves were incorporated into a yet deeper network of separating and remixing*" (Hawks and Cochran, 2006). To testify the credibility of the "mix-n-match" hypothesis as an evolutionary hypothesis, that tried to present an evolutionary narrative on the origin and evolution of the modern representative of Sapiens, Dr. Pontus Skoglund, the famous molecular geneticist of Francis Crick Institute, UK, tried to analyze the gene flows between the prehistoric, archaic human lineage, once evolved in the African continent and its descendants, the anatomically modern human species by examining the DNA samples, extracted from the prehistoric human fossils (McKie, January 5, 2020). Dr. Skoglund has designated the socio-cultural as well as socio-behavioral heritage as an important issue in his analysis, whenever he tried to track down the footprints of gene flow traversed through the evolutionary trail, once evolved in the prehistoric archaic lineages of hominins and descended to its descendants through the meshwork of braided rivers and rivulets, as the socio-cultural and socio-behavioral traits of archaic members of the hominins are either unknown or under-explored, so its inheritance from the predecessors to its successors are merely predictive (McKie, January 5, 2020). Hence, integration of issues like inheritance of socio-cultural traits along with the vertical transmission of genetic principles (that transmitted through the encrypted format of DNA) from the archaic predecessors of hominins to its present version of anatomically modern representatives of hominins is necessary to narrate the origin and evolutionary history of Sapiens, as it has rendered the formation of a diverse group of ethnic communities and their descendants of hominins in different periods in different locations. And, it has been echoed in the words of Edward O. Wilson, regarded as Father of Socio-biology and biodiversity, as he has said "*Homo sapiens have had one hell of a history! And I am speaking both of deep history-evolutionary, genetic history - and then, added on to that and interacting with it, the cultural history recorded for the past 10,000 years or so on*" (Wade, May 2, 2000). Apart from the contemporary researches on the origin and evolution of modern human species, made by molecular biologists and human genomicists, forensic anthropologists have played an important role in this field as they have noticed that the intra-molecular bond of the DNA molecule had found to be loosened in the lapse of time, which normally is found in investigations on the physically dead object of biological entities. The scientists, specifically the forensic anthropologists have also noticed that the evolutionary transcripts, those were embedded in the hard tissues of human skeleton or teeth have found to be gradually diminished due to the impact of exposition of the environmental factors like scorching sun or vortex of chilled ice from ice ages or torrential rain-fall, etc. (Latham and Miller, 2018). The scientists engaged in paleontological analysis (to determine the geological age of that fossils concerned) have empirically stated that their outcome of the investigations was found to be precise, the fossils, exhumed from the fossil sites of cold temperate, rather than the skeletons, exhumed from the fossil or burial sites in the tropical and subtropical regions of the world (Bergfeld et al, 2017; Latham and Miller, 2018). So, an endeavor of the scientists to justify the evolutionary age of the hominins, based on the DNA analysis (age dating molecular clock calibration) of any prehistoric hominin skeletons, exhumed from any tropical and sub-tropical regions are found to be remarkably different from an identical hominin skeleton, exhumed from the temperate or sub-arctic regions and such biased observation as well as so-simplified interpretation and short-sighted narratives on the origins and evolution of some archaic cousins of modern human species needs to be addressed after a critical review of the analysis, made by the evolutionary biologists and molecular geneticists.

Let's walk through the narratives of cultural inheritance as the key impetus in the feasible depiction of origin and evolution of Sapiens, which has been hypothesized by the Professor Maxime Aubert of University of Wollongong, Centre of Archaeological Science, New South Wales, Australia, one of the famous cultural anthropologists contended that in the emergence of the modern version of Sapiens, acquiring and introgression of cultural inheritance from their archaic predecessors for hundreds of thousands of years in course of vertical cycles likely rendered them to evolve as an anatomically modern representative of Sapiens in this regime of Anthropocene. Professor Aubert, along with six co-researchers from Australia and Indonesia have discovered and meticulously examined "*Pleistocene cave art from Sulawesi, Indonesia*," one of the best forms of cave art, on the Markos-Pangkep region of the Indonesian island, South Sulawesi (Aubert et al, 2014). Prof. Aubert and his research colleagues studied the intricate form of visual artwork of humans, animal figures (like pigs, buffaloes), and humanoid figures (having bird-like head and tails, etc.), had been drawn on the cave-wall, drawn by the prehistoric, indigenous communities of Sapiens and they have published their observations in the renowned science journal "Nature" in 2014. The initial reaction of the Australian and Indonesian Scientists towards this unique visual format of artwork rendered them recognize these "Asian indigenes" as an outstanding milieu of craftsmanship with a unique sense of imaginations, that would not likely be surged out of an accident, rather been acquired, and evolved as a result of a criss-cross accumulation of diverse format of cultural traits, happened as a result of migration, merging and re-emergence of derived communities Sapiens (emerged with unique socio-cultural traits), as a result of biological and cultural mix-n-match introgression of the archaic, lineages of Sapiens (likely to have distinct socio-cultural profiles in their DNAs) for ages. The uranium-thorium age-dating protocol, which has been adopted by Professor Aubert and his colleagues have finally announced that around 44,000 years ago, our indigenous predecessors left this unique piece of cultural heritage for us, that would help us out to get closer to our evolutionary root-stock, to decrypt and narrate the history of origin and evolution of humanity on this planet Earth. Dr. Peter Veth of the University of Western Australia, Australia, one of the renowned archaeologist and connoisseur of the historical arts considered the unique imaginative cave paintings in South Sulawesi, Indonesia as one of the ancient artwork from the Asian continent, historically resembled the indigene art-works prepared by the predecessors of the modern version of Sapiens in Europe (e.g. the paintings of Chauvet cave in France). The discovery of these unique imaginative cave paintings of South Sulawesi, Indonesia has intrigued Dr. Veth to state with an affirmation: "*As with the early dates of people voyaging across the sea to Australia and engaging in highly complex art, here we have [Southeast] Asian Indigenes showing human-animal relations before sapiens 'sapiens' even got to Europe*" (Greshko, December 11, 2019). Furthermore, the depiction of fundamental and imaginative ideas of Indian indigene, which has been presented in form of intricate cave artwork of Sulawesi have broken the myth, once contented by the historical art connoisseurs, as they believed earlier that Europe was finishing school for humanity as there has no such existence of unique, prehistoric art-work of Sapiens, reported from the Asian continent, before the discovery of the Cave-paintings from Sulawesi. Hence, the discovery of cave paintings in

Sulawesi has given a jolt in the mindset of the archaeologists and historical art connoisseurs, biased with "Europe was finishing school for humanity" and the discovery of Sulawesi cave paintings has inspired some contemporary archaeologists like Dr. April Nowell of the University of Victoria, Department of archaeology, Australia, to step out of the erstwhile, stereotypic vision of their seniors and firmly claimed: "*We have long known this view of Europe as a 'finishing school' is no longer tenable, and the richness of the finds from Australia and Indonesia continue to underscore this point*" (Greshko, December 11, 2019). The unique presentation of the visual arts comprised of bar-reliefs, mosaics, morals etc., on the icon cave walls, of a number of animal figures like lions, mammoths, rhinos, mammoths, humanoid figures (a composite, imaginative figures between animals and prehistoric humans) of the World Heritage Sites of Altamira, Chauvet, Lascaux in Europe have rendered the scientists to consider the archaic representatives of modern version of Sapiens as supposedly carried the creative, imaginative socio-cultural and socio-behavioral traits, originated and evolved around 40,000 YBP, and the descendants as well as migrants of these archaic representative of hominins have unlocked their abstract ideas of imagination in different parts of Europe, once contended in the DNA of their predecessors, evolved in the African continent and this evolutionary journey of cultural heritage of Sapiens have supposedly gone through in the backdrop of genetic interactions, genetic mutations, genomic introgressions of archaic representatives of Sapiens, gave rise to the anatomically modern human species on the planet Earth in this regime of Anthropocene (Marchant, January 2016; https://www.smithsonianmag.com/history/journey-oldest-cave-paintings-world-180957685/). Whether, it was the unique form of visual art, drawn by our prehistoric ancestors either on the cave-wall of Sulawesi in Asia or on the cave wall of Altamira, Europe, the unique resemblance of the cave paintings in Sulawesi and Altamira, depicts the simple but imaginative ideas of our archaic predecessors, contended the identical traits of socio-cultural and socio-behavioral inheritance, in their DNA. Dr. Elizabeth Culley of Arizona State University, Department of Archaeology, USA, and who has also been recognized as one of the experts on historical cave art has critically studied the cave paintings of South Sulawesi, Indonesia and she has claimed further: "*There's a huge variation in cultures [between Asian and European], there's a lot of space dividing these traditions...but they're also very consistent... That, to me, is the real take-home: They're contemporaneous with a very, very similar tradition, which must have some shared origin*" (Greshko, December 11, 2019). The evolutionary biologists and molecular geneticists have contended that the identical form of abstract ideas and imaginative brain-works and socio-cultural traits, either found in the archaic indigenes of the Asian continent or found among the European Indigenes, must have a common point of origin in the African continent, before taking the distinct root of migration to distinct geographical locations of Asia and Europe, wherever they had left the priceless cultural heritage for us, introduced us towards our prehistoric root of origin and evolution. So, the evolutionary biologists, molecular geneticists, and human genomicists have considered that the potential of presentation of the abstract form of art by using the brain-power with a blend of a unique sense of imagination, depicted in the cave paintings, drawn by our ancestors, would have carried that special gene in their genome and we the modern version of Sapiens, the contemporary descendants in this regime of Anthropocene, have inherited these precious traits of cultural inheritance through its vertical transmission from ancestors to its descendants in course of the evolutionary journey of hominins for hundreds of thousands years. So, we should feel proud to be recognized as the happy great-grandkids of our great-great-grandfathers and great-great-grandmothers (once lived in the African continent), bestowed with these special genes of socio-cultural cognition from them, once they have possessed these precious genes in their genomes to pass it on to their great great-grandkids and that would inevitably helped to narrate our history of origin and evolution of Sapiens in the metaphoric "Garden of Eden" around 200,000 YBP.

Hence the scientists like Chris stringer, Eleanor Scerri, John Hawks, Vanessa Hayes, and Pontus Skoglund, who have strongly supported the hypothesis on "Pan-African evolution of humanity" in their narratives of biological origin and evolution of anatomically modern human species on planet Earth, have come up with further postulations as they have contended that the anatomically modern human species would have not evolved, if it would not traverse through the labyrinth of "long-range social networking," acquired the diverse range of socio-cultural and socio-behavioral traits in their genes, and enriched its genome in course of origin and evolution of socio-cultural inheritance, accumulated through the vertical cycles of inheritance from the ancestors to their successors. So, scientists, specifically, those engaged in the exploration of the evolution of humanity on planet Earth, has contended further that the accumulation of unique genes in the genome of the modern version of Sapiens, which have been descended from their prehistoric ancestor for hundreds of thousands years, and these inherent genotypic profiles have started expressing their unique phenotypic traits in form of inheritance and transmissions of a set of morphological, genetical, behavioral, cognitive and socio-cultural traits (higher order of consciousness) like contending moral values, expression of intellects, expressing beliefs, abiding a set of socio-cultural norms, studying and analyzing potentials, expression of creative ideas, state of imagination, and symbolic thought process, etc.

Most probably the modern human species would have not evolved as modern human species on Earth if it would not have gone through the labyrinth of long-range social networking to accumulate and inherit knowledge, beliefs, set of moral values, likely to be drifted through a criss-cross meshwork of the genealogical lineages arrayed in a typical fashion of braided rivers and give rise to the origin and evolution of humanity. Logically, it's wrong to promote the hypothesis of origin and evolution of humanity in the African continent, rather it would be technically convincible to contemplate the pan-African origins and evolution of the modern version of Sapiens had likely emerged more than once, in a period (rather than a particular point of time) and it had likely taken place in a wide matrix of a continent (rather than a specific location of the continent) from the metaphoric cradle of "Eden" across the African continent and we should feel proud to acknowledge the reality that we are all descendants of African Adams and African Eves (rather than a primordial couple: Adam and Eve), had begun their evolutionary pilgrimage on planet Earth around 200,000 YBP.

References

Aubert, M., Brumm, A., Ramli, M.m Sutikna, T., Saptomo, E.W., Hakim, B., Morwood, M. J., van den Bergh, G.D., Kinsley, L. and Dosseto, A. (2014). Pleistocene cave art from Sulawesi, Indonesia. *Nature* 514: 223–227. https://doi.org/10.1038/nature13422

Barras, C. (March 3, 2016). The father of all men is 340,000 years old. https://www.newscientist.com/article/dn23240-the-father-of-all-men-is-340000-years-old/

Bergfeld, A.K., Lawrence, R., Diaz, S.L., and Pearce, O.M.T. (2017). N - glycolyl groups of nonhuman chondroitin sulfates survive in ancient fossils. *PNAS.* 114(39): 201706306. DOI:10.1073/pnas.1706306114

Cann, R. L. (2013). "Genetics. Y weigh in again on modern humans." *Science.* 341 (6145): 465–67.

Elhaik, E., Tatarinova, T. V., Klyosov, A. A., and Graur, D. (2014). The 'extremely ancient' chromosome that isn't: a forensic bioinformatic investigation of Albert Perry's X-degenerate portion of the Y chromosome. Eur J Hum Genet. 2014 Sep;22(9):1111-6. doi: 10.1038/ejhg.2013.303. Epub 2014 Jan 22. PMID: 24448544; PMCID: PMC4135414.

Greshko, M. (December 11, 2019). Ancient cave art may depict the world's hunting scene. https://www.nationalgeographic.com/science/article /ancient-cave-art-in-indonesia-may-be-worlds-oldest-hunting-scene.

Grine, F. E., Henshilwood, C. S. and Sealy, J. C. (2000). Human remains from Blombos Cave, South Africa: (1997–1998 excavations). *J. Hum. Evol.* 38: 755–765.

Hawks, J., and Cochran, G. (2006). Dynamics of adaptive introgression from archaic to modern humans. *PaleoAnthropology*. 2006: 101-115.

Johanson, D. C. and Wong, K. (2009). *Lucy's Legacy*. New York, NY: Harmony Books.

Latham, K. E. and Miller, J. M. (2019). DNA recovery and analysis from skeletal material in modern forensic contexts. *Forensic Sciences Research.* 4:1, 51-59, DOI: 10.1080/20961790.2018.1515594.

McKie, R. (January 5,2020). The Search for Eden: In pursuit of humanity's origin. https://www.Theguardian.com/world/2020/jan/05/the-search-for-eden-in-pursuit-of-humanitys-origins.

Mendez, F. L., Krahn, T., Schrack, B., Krahn, A. M., Veeramah, K. R., Woerner, A. E., Fomine, F.L., Bradman, N., Thomas, M.G., Karafet, T. M., and Hammer, M. F.(2013). An African American paternal lineage adds an extremely ancient root to the human Y chromosome phylogenetic tree. *Am J Hum Genet.* 92(3):454-9. doi: 10.1016/j.ajhg.2013.02.002.

Wade, N. (May 2, 2000). The Human Family Tree: 10 Adams and 18 Eves. https://www.Nytimes.com/2000/05/02/science/the-human-family-tree-10-adams-and-18-eves.html.

www.ingramcontent.com/pod-product-compliance
Lightning Source LLC
LaVergne TN
LVHW010718210926
848485LV00008B/1442/J